Author's Biography

Peter Farwell is a Chartered Professional Accountant and Certified Financial Analyst.

He is uniquely positioned to write The Personal Computer, Past, Present and Future. Peter was an early purchaser of an Apple II and an avid user of VisiCalc.

He is a retired partner of Public Accounting Firm Ernst & Young. He was the leader of the Canadian Firm's services to the High Technology Industry for fourteen years.

Peter was the co-author of several studies of the Canadian High Technology Industry. These included a study of trends in the Canadian Software Industry, conducted by interviewing 12 of the CEO's of Canada's leading Software companies. He coordinated the Canadian Electronic Industry's participation in a four country, four industry study of Total Quality Management practices.

Peter has written articles and given speeches on aspects of Strategic Planning and Financing for High Technology companies. These included a lecture to the Association of Canadian Venture Capital Companies on the six stages of growth of technology companies, based on a 1972 landmark paper on the subject by Professor Greiner of Harvard University.

Recently, in 2012 and 2013, he has coauthored three studies of Research In Motion that endeavor to determine its chances of survival and what changes management have to make to do so.

Introduction

This is a book about the Personal Computer, its Past from 1975 to 2011, its Present from 2011 to 2017 and its Future from 2018.

The Past

The Personal Computer's past is taken primarily from my previous book: A Short History of the Personal Computer, published in 2013 and available on Amazon.com. It focusses on the growth in global sales volume from its inception in 1975 to the peak in its sales volume in 2011. In A Short History of the Personal Computer, our story stressed the importance of the sequence of Applications Software created for the PC. We believe it was this sequence of software innovations that was the key driver of the popularity of the PC, and led it to increasing sales, time and again, over the PC's 36-year history from 1975 to 2011.

However, none of this success would have been possible without two linked series of computer innovations that provided the platform for the series of application software that we have highlighted in this book: (1) The developments of the microprocessor hardware, led by Noyce, Moore and Groves at Intel; and (2) The development of the PC Operating Systems, led by Bill Gates and Paul Allen at Microsoft.

Because of the Importance of these innovations to the success of the Personal Computer, we have included a review of the history of both the PC hardware and the PC software in an Appendix, Appendix F.

The Present

A Short History of the Personal Computer was published using Amazon's Create Space publishing system in the fall of 2013. At that time, we concluded by observing that in 2011 and 2012 there was evidence of a peaking in global PC sales in 2011 at 365 million units. In 2017, we can see that this peaking has occurred and indeed we are now in a period of precipitous decline in global PC Sales.

We have added a section describing the continuity of PC global Sales from 2007 to 2017. It includes some elaboration on comments in A Short

History of the Personal Computer on the causes of the decline in PC sales after 2011, which we attribute to (1) the arrival of the Smartphone that took away much of the PC's function as the primary digital communication device for internet users; (2) the lengthening of the PC replacement cycle, and (3) The lack of another mega App. In this book we elaborate on these causes of the PC's decline.

The Future

Looking to the future, we explore possible sources of renewed growth in overall PC sales. We conclude with a list of positive things to watch for that might indicate a halting of the slide in PC sales and growth in the future.

Contents

The Past

Introduction to the Past History of the Personal Computer

This story begins in 1975 and is still ongoing, although it is currently being overshadowed by the Smartphone story. The development of the PC from the foundation of "The Homebrew Computer Club" in Menlo Park, Silicon Valley, California, to the creation of the Apple II, the introduction of the IBM PC, the leadership of Dell Computers, the arrival of the Internet and the production of the low-cost models of Acer, Asus and Lenovo.

It makes the case that the growth of PC sales has been attributable not to the hardware developments, as remarkable as they were, as much as to the software developed for the PC. Key software came along at just the right times to take the use of the PC to rapid growth, again and again.

In 2012, the growth curve for the PC was flattening out. This essay attributes this largely to the creation of another Disruptive Technology (to use the term made famous by Clayton Christensen), the Smartphone first introduced in 2004 by Research in Motion in Waterloo, Canada, and now being made pervasive worldwide by Apple and Google.

Apple Computers

The Apple 1

This history of the PC begins with the formation of the Homebrew Computer Club. It was founded by Fred Moore and Gordon French on March 5, 1975, in the heart of what we now refer to as Silicon Valley, California, as a group of people interested in building their own computers.[1] The group included Bill Gates and Paul Allen, and most notably for our purposes, Steve Wozniak and Steve Jobs.

Coming out of that first meeting, Wozniak had a brainwave; a self-contained personal computer built around a microprocessor, the computer on a chip being produced by Intel and others about this same

[1] Walter Isaacson's biography of Steve Jobs, page 60

time.[2] Wozniak actually chose a chip made by MOS technologies as the core of his PC, because it was cheaper than Intel's.

Shortly thereafter, the Apple I made its debut. This was a PC built for hobbyists. It was in fact just a printed circuit board built around the microprocessor. It had no power supply, monitor or keyboard.[3] Purchasers had to supply these and plug them in. There were no application programs. Purchasers had to do their own programming using a version of the Basic

Program Language written by Wozniak especially for the Apple I. Fortunately, enough of them (about 200) purchased the circuit boards to get the Apple Computer Company off the ground. It had lots of competition, from Altair for one, and was used only by a small minority of the Homebrew club members. However, it was enough to provide a base for the development of "The Next Big Thing", the Apple II.

---

[2] Intel had launched its 8080 microprocessor, considered by many to be the first general purpose microprocessor with 4500 transistors on it, in 1974, per the Intel website history line. 3 Isaacson's biography of Jobs, page 68.

[3] Isaacson's biography of Steve Jobs, page 68.

The Apple II

The Apple II "was a phenomenal improvement" over the Apple I.[4] Steve Wozniak had been working on a successor almost from the time the Apple I had been completed. The Apple II, as they named it, would come in its own molded plastic case, and contain its own power supply, keyboard and monitor. It would produce high resolution graphics in color. As Wozniak envisage it "this would be a matter of efficiency and elegance of Design"[5]. The Apple II would be designed around text and graphics that would be embedded in the Apple II's own DRAM memory system. It was much faster and employed fewer chips in the printed circuit board than its predecessor, and so was smaller and cheaper to make, and it was the first personal computer to start up ready to use with

its Basic Operating system burned into its ROM (Read only permanent internal memory), rather than having to be "booted up" by the user with a tedious process. It was designed for the average home user, rather than

[4] iWoz chapter 13.
[5] ibid

the hobbyists of the Homebrew Club.

Each of these advances, incorporated in the Apple II, was a marvel itself. For example, the power supply was extremely innovative. It was developed for the Apple II by Rod Holt, an acquaintance of Steve Jobs. Jobs wanted to have a power supply that didn't need a loud fan to cool it. Holt built one that switched itself on and off very fast, so that it could operate without heating up much, unlike prior power supplies for computers. Therefore, it required no noisy fan for cooling. In Jobs words "that power supply was as revolutionary as the Apple II logic board"[6].

VisiCalc

As marvelous and innovative as each of these advances incorporated in the Apple II were, none was the real key to its success. The key was a software program which ran on the Apple II called VisiCalc. VisiCalc was the first Spreadsheet program for personal computers.[7] It was developed by Dan Bricklin and Bob

---

[6] Walter Isaacson, 2011. Steve Jobs, ISBN 978-1451648539, page 74.
[7] https//en.wikipedia.org/wiki/VisiCalc

Frankston, founders of Software Arts, and distributed in 1979 for the Apple II. It was the first fully interactive "row and column" financial calculation software developed for a personal computer.9 It propelled the Apple II from being a hobbyist's toy to a useful tool for business, two years before the introduction of the IBM PC.[8]

To understand why it was the key to Apple II's success, let's go back into history, my history with actuarial computations.

In the summer of 1957, I joined Confederation Life, ("Confed"), a now defunct Canadian Insurance company, as an actuarial summer student. One of my jobs that summer was to gather the data on the most recent year's Canadian mortality records, assembled by hand from inputs from Canadian insurance companies, and to integrate these with the accumulated data from prior years. The result was a new Canadian mortality table, right up to date, that could be used for calculating insurance premiums and the company's insurance liability reserves. This work, as I recall, was done using NCR calculating machines

---

[8] http://arstechnica.com/business/2012/08/from altair-to-ipad-35-years-of personal computer market share

to do the calculations that were then entered on massive and multitudinous hand- written spreadsheets. It took most of the summer.

The next summer found me back at Confed with the same updating task, but this time using IBM punch card machines, calculators, sorters, accumulators, and such. This was a little faster, but still took most of the summer as I learned how to use the machines and how to feed in the punched cards without destroying the data.

The third summer I was back again. This time Confed had purchased one of the first IBM 705's. The IBM 705, was what we now call a mainframe computer. It was massive, a machine of tubes that could be switched on and off to signal and store information in the binary integer system. In that system each bit is an electric tube switch that was either on, signifying a "1", or off, signifying a "0", — machine language; Eight Bits were combined into a Byte that could represent, in the binary system, 256 different data points; in the ASCII language bytes were used to represent the letters of the alphabet, in both lower and uppercase, integers, punctuation marks and mathematical operations such as Add and Subtract; The ASCII language could in turn be used to express any sentence or logical operation. (For a fuller description of the ASCII language and its place in the computer software hierarchy, please see Appendix B, A hierarchy of Computer Hardware and Software)

The IBM 705, a significant innovation in its time, was constructed before the advent of the transistor; it generated a huge amount of heat and required an immense cooling system, such that the whole machine took up an entire floor in the Confed building.

My task that summer was to write, in an assembler programming language, a program that would do the same mortality table update that I had done the preceding two years. As you can imagine it would have resulted in very large savings in the effort required to update the mortality tables and it would have been repeatable each year.

The point of this diversion is that a personal computer that could run an electronic spreadsheet would constitute a giant leap forward from the manual creation of spreadsheets. The calculations could be done more easily, corrected more easily, the data could be stored indefinitely, or printed numerous times, and the work could be repeated time and time again in exactly the same way.

This was extremely valuable to actuaries building mortality tables and insurance liability schedules, to accountants creating depreciation and other schedules required to produce audited financial statements, to engineers building tables of specifications for new products and services, to investment analysts tracking stock and portfolio performance and to many others who required spreadsheet data to do their work. VisiCalc was the first program that did this with the convenience of the personal computer. Each possessor of a personal computer could finally develop and run his or her own personal spreadsheets while their compatriots were doing theirs without tying up the company's mainframe(s). The result was a giant leap forward in productivity and ease of use. So once people found out what they could do with the Apple II, they flocked to buy them, and sales took off.

In the world of the IBM PC, described later, VisiCalc was followed by Lotus 123, the dominant Spreadsheet program in the PC DOS world, and still later by EXCEL in the Microsoft Windows world.

There are two other important pieces of the Apple II puzzle that I would like to address before moving on to the next phase in the PC story.

Mike Markkula's contributions

The first of these is the input from Mike Markkula. Markkula had worked at Intel and made millions on stock options before retiring at about age 30.[9] He had kept involved in the technology Startup field as an angel investor and consultant who excelled in strategy, organization and marketing. Jobs was introduced to him by a venture capitalist and they hit it off.

To start with, Markkula insisted that they write a Business Plan for the Apple Computer Corporation and he agreed to invest if he liked the plan. He ended up providing the

[9] Isaacson's biography of Steve Jobs, page 78

company with $250,000 for a one third interest. This was more than enough funding to launch the Apple II.

Even more importantly, Markkula took over the marketing function for Apple. He wrote his approach to marketing in a one-page memo "The Apple Marketing Philosophy". It set out three principles that are so good they bear repeating here as a guide for any high-tech company.

Exude Empathy – Make every effort to understand the needs of your target customers as those customers perceive them. We will come back to this principle later when we discuss the operating approach of Dell Computers. But the key point is that those needs should drive everything the company does.

Focus - Eliminate all the lesser opportunities and put all the company's resources into pursuing its best opportunities, those required to enable it to succeed.

This is extremely difficult for an entrepreneur to do but absolutely key to success.

Impute - This awkward word was used to convey the need to infuse the company's entire operations with the same values as its main products or services. This would allow customers and other stakeholders to understand what the company stood for from looking at any part of its operations. If you are introducing an innovative product or service, you must market it in an equally innovative way. This is how you will build trust in your ability to innovate. This is a mantra that has guided Apple's marketing, particularly product launches, to this day.

Markkula's investment of $250,000 was critical but his marketing practices were even more significant.

Wozniak's Graphics Capability

The other piece of the puzzle is the graphics capability Wozniak built in to the Apple II. It was far superior to that in any other PC at that time and gave Apple a significant competitive advantage. It helped make the APPLE II fun and easy to use. As Wozniak tells it in his auto-biography iWoz,[10] he designed the Apple II with a colour graphics capability coded right into its main ROM memory. The Apple II was designed to work with any TV as a monitor. It had game control Paddles that could be plugged in and it had sound. It was also a high-resolution machine: "You could program every single pixel on the screen."[11] And it had more CPU memory than its competitors, 48k to start with, and 64k in later versions. (The latter as much as that IBM 705 mainframe that I had to program at Confed in 1959, seventeen years earlier.)

All these features made the Apple II attractive to creators of computer games. As a result, a whole new community of start-ups was formed to write software and build

[10] iWoz, chapter 13

[11] iWoz,chapter 13

attachments that could run computer games on the Apple II. This was a very competitive market and it was hard for any company to build much traction in the early computer game market.

One of the most popular games was Pong,[12] which enabled a player with a paddle to hit a bouncing Ping Pong ball against a wall. Wozniak wrote a version of this for the Apple II.

However, Wozniak noted that, it was VisiCalc that caused the sales of the Apple II to explode, changing the market for personal computers from that of hobbyists playing games to business users who achieved a quantum leap in productivity. “After a couple of months, the business people were something like 90% of the market” he wrote[13]. Ultimately the Apple II, in its various versions (Apple II+, Apple IIe) sold over 16 million units and put the Apple Computer Company in the Fortune 500 and right in the middle of one of the greatest technology revolutions ever.

---

[12] Toronto Globe and Mail,2013: “Atari, which celebrated its 40th birthday (in 2012), was founded in 1972 by two American engineers, Nolan Bushnell and Ted Dabney. Originally intended as a video game software firm and responsible for hits such as Pong – Atari eventually achieved some of its greatest success in the (PC) hardware market.” Atari was an early leader in the PC market, before the APPLE II reached its prime, because of its success in the new entertainment media of computer games. Atari, unfortunately, went bankrupt in early 2013

[13] iWoz, page 220

The IBM PC

The IBM PC was launched on August 21, 1981 and in due course ushered in a new era for the Personal Computer. However, compared to the take up of the APPLE II and its close connection to the use of VisiCalc, the IBM PC story is much more complex, and took place over more than a decade after its launch in 1981. It is not so much a story of the sales of IBM's PC's as a story of the sales of its clones. The software for the IBM PC, in particular the Operating System and the Word Processing software for it, went through a succession of iterations, which are not well documented, before the IBM PC and Clones reached dominance with about 90% of the PC market in 1992, eleven years after it was launched.[14]

While is not possible to show the close correlation between the sales of the IBM PC and its Word Processing software that were present between the sales of the APPLE II and VisiCalc, there are some clues to the importance of the Word Processing software to the take up

[14] http://origin.arstechnica/articles:from Altair to iPad: 35 years of personal computer market share; Jeremy Reimer, august 2012

of the IBM PC, particularly in the office.

The story unfolded as outlined in the following paragraphs.

IBM's Personal Computer series

The IBM PC was by no means IBM's first attempt at a Personal Computer, but it was the first truly stand-alone unit.

By 1981, IBM had produced and marketed a series of units that it referred to as Personal Computers. So perhaps to IBM at the time, the computer that became known as the IBM PC was just another in the series- IBM referred to it as the model 5150.[15] As we shall see, it was different!

First, let's look at the IBM series of personal computers. According to the IBM archives, the origin of a single user computer goes back to 1973. In that year, IBM produced a working prototype of its 5100 "Portable Computer" that allowed a user to perform desktop calculating operations, but not with the spreadsheet capability of VisiCalc, and a limited variety of "canned"

[15] Wikipedia: The IBM PC

applications, which presumably did not include word processing. This limited capability machine was referred to in PC magazine in 1983, looking back 10 years, as a “revolutionary concept” because of its single user capability.

The IBM 5100 was finally available for sale in 1975. It weighed 50 pounds and cost about $9,000. It was aimed at “engineers, analysts, statisticians and other (mathematical) problem solvers”; but again, without the capability, ease of use and convenience of a VisiCalc spreadsheet program. It was withdrawn from the market in 1982.

The IBM 5110 was introduced in 1978. It was offered as a “full function computer” aimed at “all business and industry”. It added a number of accounting functions, such as a general ledger and accounts payable applications and could be used to generate reports based on these. It had a keyboard, CPU and display screen, and could access magnetic tape or diskette storage, and could drive an IBM printer. It too was withdrawn in 1982.

In 1979 IBM introduced the IBM 5520, a computer system that

provided “advanced text processing and electronic document distribution.” This was not a personal computer so much as a computer system aimed at the production, editing, storage and distribution of documents such as correspondence and manuals that are a big part of the daily routine of a business office. Each typist had a personal terminal or “display station”. So, they were getting closer to the personal computer functions as we know them today, but still were not quite there, either in terms of providing single user independence, in terms of the variety of capabilities available to each single user, or in terms of the single user price range. One suspects that the versatility of the Word Processing software available was also limited.

In 1980, IBM introduced the Displaywriter, a “low-cost” desktop text processing system, at a time when “most documents were still created on manual or electric typewriters” with no memory. This system allowed text indentation, right margin justification, centre and underscore functions, spelling check, and document storage and recall for editing – all features that were novel at that time. A single user system cost $7,895. A system of three displays sharing a printer and a paper handler sold for $26,185.

"The IBM PC"

The IBM PC launched in August of 1981 was different from this series described in the IBM archives, that we have just outlined. Before reading this section, you might find it useful to review Appendix B, A Hierarchy of Computer Hardware and Software, for background information.

IBM had been fretting for some time about the creation of a microcomputer for a single user that could operate on a standalone basis and allow a user to perform a multitude of tasks. Other computer manufacturers, such as Apple, Commodore and Atari, were selling such machines into the business world and eating into IBM's primary market. But IBM had been concerned that such a machine produced by IBM would cannibalize its existing array of business machines, and so delayed its own development of a true Personal Computer. The growing success of Apple with the aid of VisiCalc forced IBM to change its tune.

To short circuit the development period, IBM decided to build the IBM PC on the Intel 8080 series of microprocessors, rather than one of its own, and to hire Microsoft to produce the computer's operating system. Further it decided to make its PC design open so that other manufacturers could clone it without peril. Both these decisions were atypical for IBM, and yet each contributed to the success of the IBM PC in the market place.[16]

The group within IBM charged with creating the IBM PC had talked to Gary Kildall, the founder of Digital Research, about using the CP/M Operating system that he had developed and had become the microcomputer.industry standard by 1980.[17] However, they were unable to reach an agreement and IBM decided to go with Microsoft to provide both the Basic programming language and the Operating System for the IBM PC; the latter something Microsoft had never done before.

Seattle Computer Products was producing an Intel 8086 microcomputer and had hired Tom Patterson to create an operating

---

[16] Wikipedia on Wintel and other sources. Wikipedia notes: "IBM published the technical specifications and schematics of the PC, which allowed third-party companies to produce compatible hardware, the so-called open architecture."

[17] Penguicon .sourceforge.net: a short history of CP/M. In 1974, working for Intel, Gary Kildall used his programming language PL/M to create CP/M, the first Operating System for microcomputers, to run on the Intel 8080 microprocessor. However, Intel allowed Gary to take the rights to CP/M and in 1976, he left Intel to start a new company that became Digital Research to sell the Operating System.

system for it. When Digital Research was slow coming out with the 8086 version of CP/M, Patterson wrote a system with the "Look-and-feel of CP/M called QDOS". Seattle Computer showed it to Microsoft late in 1980 and Microsoft bought the rights to QDOS for $50,000. Microsoft then hired Patterson to convert QDOS to run on the IBM PC under the name PC DOS. It also retained the right to sell copies of the system to other PC manufacturers under the name MS-DOS.[18] This Operating System eventually became the Micro Computer industry standard until it was replaced by Microsoft's Windows in 1989.[19]

"Other manufacturers soon reverse engineered the IBM PC BIOS to produce their own non-infringing functional copies. Columbia Data Products introduced the first IBM-PC compatible computer in June 1982. In November 1982, Compaq Computer Corporation announced the Compaq Portable, the first portable IBM PC compatible. The first models were shipped in March 1983".[20] Soon the clones were outselling

---

[18] Penguicon.sourceforge.net

[19] Wikipedia/ Microsoft Windows: Microsoft Windows version 3.0, released in 1990, was the first Microsoft Windows version to achieve broad commercial success, selling 2 million copies in the first six months. It featured improvements to the user interface and to multitasking capabilities.

20 Wikipedia/IBM PC

IBM's own by a wide margin, principally by producing a better machine and selling it at a cheaper price than IBM. However, IBM was the market leader in the 1980's.

Word Processing Software

The IBM PC (including clones) was not the instant success that we have come to expect in the Smartphone and Tablet computer markets in 2012. In the first year, market acceptance was slow in spite of IBM's longstanding leadership in serving the business market. In large part, this was deliberate. IBM did not want to have the PC take sales from its existing array of computer equipment for the business office.[21] For example, IBM chose the EasyWriter as the word processor for the IBM PC.[22]

The EasyWriter

The EasyWriter had been written by John Draper, initially for the Apple II, probably in 1977. It was a simple word processor. The EasyWriter Professional launched in 1980, competing against Wang, DEC and IBM. EasyWriter "was the only package, at the time, to display text

---

[21] IEEE Annals of the History of Computing: October-December 2006? Thomas Bergin/ The origins of Word Processing Software for Personal Computers, p 37
[22] ibid

on the screen exactly as it would appear on the printed page".[23] (perhaps the first Word Processor to achieve WYSIWYG, "what you see is what you get"). Personal Computing magazine judged it the best of the bunch in January 1981. However, it had some serious limitations including the fact that it used a non-standard disk operating system. Nevertheless, when IBM was unable to cut a deal with Seymour Rubinstein of MicroPro, (see below) it opted to go with EasyWriter, even though the version to run on the IBM PC was not ready and they had to use an earlier version. This turned into a disaster. A typical commentary was a PC magazine article reviewing it titled "The Not So Easy Writer".

WordStar

At this time (1981), the dominant Word Processor was MicroPro's WordStar. WordStar ran on the CP/M operating system, the industry leader in 1978, and became the first of a series of three serial monopolies of Word Processing software.[24] Rubinstein established MicroPro in 1978 and teamed up with a CP/M programmer John

[23] Ibid,p.36
[24] Ibid, p. 32

Barnaby, who wrote the program for WordStar. It incorporated the best features of existing WP software running on minicomputers but ran on personal computers using the CP/M Operating system.[25]

It was Rubinstein's goal to address the Word Processing market "to get a larger market" as the lead part of a software suite aimed at the growing micro computer industry.

WordStar launched in June of 1979, with a lot of innovative features, and over the next five years became the runaway leader in Word Processing software, hitting its peak in 1984. A key note of particular interest to us is the following observation in the IEEE article:

"Rubinstein believes that WordStar provided a reason for someone ...to buy a (microcomputer), because WordStar ... made the computer functional and useful immediately." To Rubinstein, WordStar was the first "killer application" because of its dominance of the marketplace and its millions of dollars in sales."[26]

---

[25] Ibid, p. 39

[26] IEEE article, p.39

However, MicroPro and WordStar fell on hard times as a result of several events in 1984, the most serious of which was a heart attack suffered by Rubinstein. Its place as the industry leader was taken by the second serial monopolist, WordPerfect, running on PC- DOS and MS-DOS.

WordPerfect

WordPerfect's was the leading Word Processing software for the period from 1984 to about 1990.

WordPerfect was originally written for a Data General Minicomputer in 1979. The authors incorporated a company, Satellite Systems International and sold it under the name WordPerfect. A version that ran on PC DOS was produced in1982.The software was a clear improvement on WordStar, with automatic paragraph numbering, and automatic numbering and placement of footnotes.[27]

In 1989, WordPerfect 5.1 for PC DOS was released with even more useful features. These included Macintosh

[27] Wikipedia on WordPerfect

style pull- down menus and support for tables.

"Some would argue that this version of WordPerfect is still the best Word Processor in the Business".[28]

While all this was going on in the PC software markets, sales of the IBM PC, after a slow start, were booming. In 1986 sales of the IBM PC (including clones) reached 50% of total PC market sales, and in 1988 represented 50% of the installed base of PCs.[29] However the bulk of these sales were IBM PCs made by the clone manufacturers, such as COMPAC, using the strategy of making a compatible PC that had better features and was faster than the IBM product and selling it more cheaply, both of which were easy to do because of IBM's concerns about making a PC that would cannibalize sales of existing IBM products aimed at the Office market.

---

[28] Ibid

[29] IEEE; Thomas Bergin's article, p.43

The IBM PC/2

IBM was the biggest vendor of IBM PCs, but it could see that it was losing even larger sales to the clone makers; and it decided to do something about it.

In 1987 IBM launched the PC/2. The PC/2 featured a new hardware platform, while running the same software as the IBM PC. One of the new features was a Micro Channel architecture bus that permitted faster data transfers within the PC itself. This bus replaced the open AT bus used on the IBM PCs. One significant result was that the PC/2 would not accept any of the add-in cards that had been produced for the IBM PC. As a result, these would all have to be redesigned by their manufacturers, a costly and time-consuming process.

In addition, IBM made it virtually impossible for the PC/2 to be cloned. As an alternative for clone manufacturers, IBM kindly offered to license the PC/2 provided the licensee paid a royalty for each PC/2 compatible machine that it produced and, on all IBM,-compatible machines that the licensee had manufactured in the past. While some of the clone makers signed such a license, an important group refused to do so. This group agreed to create an improved open bus type, the new EISA bus that would be available to any clone maker and retain backward compatibility with the IBM PC bus.

These decisions dramatically altered the dynamics of the PC market.[30] IBM was no longer in control of the bulk of the market. 23 IBM had planned to replace PC/DOS with a much-improved OS/2 operating system. It had been working with Microsoft to produce this.

While OS/2 was designed to be compatible with PC/ DOS, unlike the PC/2 hardware, Microsoft decided to produce its own new Operating System, Windows, and to encourage the clone makers to use it instead. Windows for the IBM PC and clones was launched in 1989. In due course, Windows became the dominant Operating System for PCs. By 1996 Microsoft had no significant competitors in the PC Operating System Market.[31] As a result, the PC/2 strategy turned into a bit of a disaster for IBM. The market power

---

[30] Wikipedia: Wintel

[31] Ibid

that it had sought with the PC/2 strategy passed instead to Microsoft and has not been relinquished to this day (2013).

Meanwhile, parallel developments were occurring in the Word Processing Software market described above. The Word Processing Software market, had initially been a very competitive market with many participants; but, as noted above, it came to be dominated in turn by three "serial monopolists".

Word

In 1989, Microsoft launched Windows for the IBM PC and Clones market and along with-it Word for Windows (the third and present WP serial Monopolist). As noted above, Windows became the dominant Operating System for the IBM PC and Clones and by 1996 had achieved a market share of close to 90%. Microsoft's monopoly power in the PC marketplace was undoubtedly a key factor in the runaway success of Windows 1989 and later versions. But it is interesting to note that this coincides with the success of Word in this marketplace at about the same time. Each software system went through a number of iterations before taking off, but these happened at roughly the same times. It is our thesis that this was more than a coincidence and that the arrival of a solid WP package,

Word, contributed to the strength of Windows and was a major contributor to the rapid growth of the PC market that occurred after 1989.

Why was Word Processing so important to the popularity of the Personal Computer? It may seem like a silly question today when PCs are so pervasive – everyone has one, many more than one. But in the 1970's and 1980's, it was a question technology people puzzled over.

Apple had shown that PCs could be useful as more than a hobbyist's toy or a machine to play simple games on, with the help of a colour graphics capability and the VisiCalc software that broadened the user market to include spreadsheet users: accountants, engineers, architects, sales and marketing people and the like.

The IBM PC was targeted at a much bigger market, the office market. To make the sales in that market, IBM and clone manufacturers touted the improvement in efficiencies that would be possible. Many companies took that for granted and bought tons of the machines, but in fact it was over a decade after the launch of the IBM PC that most companies began to realize the major productivity improvements that the suppliers had foreseen. This sea change in productivity occurred around the typing function.

In the old days (up to 1980), letters, documents and the like were produced by a process that involved the author dictating the document either to a secretary who took shorthand or at a later stage, to a tape recorder; the secretary, or typing pool, took the dictation and typed it up on a variety of typewriters that became more sophisticated as time passed; the typed document was returned to the author who corrected, changed it, and returned it to the typist who corrected the original by a process of cut and paste, white outs and retypes, or in severe cases retyping the entire document; after another round of proof reading, the document was printed, circulated and filed. Believe it or not, this tedious process was still being followed by many companies into the late 1980's and in some cases the early 1990's. Finally, after the arrival of decent operating systems, PC DOS followed by Windows, and decent word processing software, WordPerfect, followed by Word, companies began to change the process of creating documents.

A new generation of authors learned to type their own documents on their PCs, do their own corrections and initiate their own printing, circulation and storage. The job of the secretary was changed drastically as typing was no longer their main function- their numbers were greatly reduced, and their title was changed to Administrative Assistant. Where many authors had their own secretary under the old process, the great majority came to share Administrative Assistants, who could now look after several authors. Thus, the earlier efficiencies in the office market, touted by the PC manufacturers, were finally being realized. But in many cases, it was in the early 1990's that this move forward occurred. And of course, it was the Word Processing Software, WordPerfect running on DOS, and from 1989 Microsoft's Word running on Windows, that made this possible. Over the 10 plus years after 1981, the PC became pervasive in the office market.

The WP software also contributed significantly to the growth of PCs in the home market, where as Seymour Rubenstein foresaw it gave the man or woman in the home, something really useful and general purpose to do with the PC, something that the WP software made easy to do; type letters and documents.

In summary, just as VisiCalc spreadsheet software led to a large expansion in the market for PCs in the early 1980's, Word Processing software was a major contributor to the next large expansion in the PC market, both in the office and in the home, that occurred in the late 1980's and early 1990's. Of course, there was much other software created for the PC market; for example, database programs, particularly relational database managers popularized by Oracle and Terra Data, presentation programs such as PowerPoint, and many others. But each of these was aimed at and used by only a part of the office market. The really pervasive use was Word Processing.

The Graphic User Interface (GUI)

A history of the personal computer would not be complete without a reference to the GUI, since it amounted to as much of a revolution in the way we use computers as any other innovation. In essence, a GUI is software that allows a user to display and manipulate graphic objects on the computer and to use these manipulations as the main way of operating the computer, usually by means of a "mouse". Previously, Operating Systems display typed lines of text as "Command Line Interfaces" (CLI's), the system used by MS DOS and all predecessors.

While the history of the GUI can be traced farther back, we will start our history with the work at Xerox Palo Alto Research Center (PARC). One of PARC's goals was the "humanizing of computers". It developed the first usable GUI for its Alto computer in 1974. Steve Jobs was familiar with the work at PARC and discovered that the GUI was not patent protected. Steve decided to incorporate it in his next line of computers, the Macintosh line, released in 1984. APPLE included a number of improvements in its version of the GUI, including developing overlapping windows (a portion of the computer's monitor screen that operated independently from the rest of the screen), icons (graphic images) that could be used to initiate actions by the computer, a fixed menu bar, drop down menus, and a trash can.

These innovations were quickly adopted by most other computer Operating Systems developers and in many other devices as well. Microsoft incorporated the GUI in its first version of the Windows Operating System, Windows 1.0, in 1985. Microsoft gradually improved its version of the GUI and by 1995, in Windows 95, it was a high-quality offering.[32]

We have not otherwise commented on the Macintosh, because, while it was an immense success for APPLE, by the time it was introduced, the IBM PC (and clones) had taken over

[32] http//www.Linfo.org/gui.html

the personal computer market. But we will show the power of the use of graphics by showing our short history in graphic form against a backdrop of the history of computer unit sales in two graphics. The first, Appendix A, shows the history of the first 20 years of the personal computer from its inception in 1975 to 1994, the Pre-Internet Era.[33] The second graph, Appendix C, shows the history from the first commercial use of the Internet to 2012, The Internet Era.[34].

[33] From data extracted from Arstechnica article
[34] From data obtained from Gartner Group surveys displayed on Wikipedia

The Next Great Leap Forward

This sets the scene for the next great leap forward, the conversion of the PC from being primarily a workstation, to being primarily a communication device. The driver in this case was the arrival of the Internet and the World Wide Web. We are going to tell this part of the story by looking at Dell Computers, now Dell Inc, ("Dell"), since Dell was both an instrument of the growth in PC's, that occurred as a result of the arrival of the Internet, and one of the principal beneficiaries.

Dell Computers

I first encountered Michael Dell at a High Tech conference in San Francisco, in 1990. It was a well established annual two-day event at which young technology companies, including start ups, hoping to attract funding or partnerships, would show their new ideas, products and services to an audience of venture capitalists and established tech companies. The conference also had some luminaries who presented a series of keynote speeches. Michael Dell was one of the latter, slated as the keynote speaker before lunch on the first day.

Dell's subject that year was : What's wrong with the Personal Computer? He was assigned a time slot with about 45 minutes and a half hour question period to follow. Dell delivered 8 zingers on what was wrong with the PC industry; each one representing a significant business opportunity; and he delivered them all in eight minutes and sat down. The audience was so stunned that there were few questions and the conference organizers had to start the lunch break about an hour early.

As for me, on returning home and doing a bit of research on Dell Computers, I bought a few shares of the company. Dell had gone public in 1988 and had immediate success in selling PC's, but this success was not showing up much on the bottom line and so the company's shares drifted along without much movement either way until about 1992. In that year, the price moved up a bit and Dell did a three for two share split while maintaining its price. The share price stumbled along at the new level for several more years and then took off like a rocket in 1995. Between 1992 and

1999, the stock was split two for one six times, 96 times the stock at Dell's IPO, and in 2000 the split stock hit an all time high of $58 per split share.

How did Michael Dell and company generate all this wealth? Dell started the business to manufacture and sell Personal Computers in innovative ways, when he was still at school in 1984. The business experienced some success and in 1988 it went public, raised $30,000,000 and attained a market value of $85,000,000.

By that time, Dell was well known for its distinctive direct sales approach. Dell contacted and sold Personal Computers directly to its customers, not using the established retail and wholesale channels employed by others. This approach enabled it to learn quickly exactly what its customers wanted in their Personal Computers, and to modify what it was selling to meet these needs more quickly than any other PC vendor. (Incidentally, this is still the company's mantra to this day, as illustrated by this quote from the home page of the Dell website: "From unconventional PC startup to global technology leader, the common thread in Dell's heritage is an unwavering commitment to the customer".)

But the whole story of Dell's success goes much deeper than this direct contact with customers.

Dell's management philosophy is grounded in the Total Quality Management (TQM) system, whose principle originator was W. Edwards Deming, an American Physicist who developed the methodology that became known as TQM in the 1930's, 40's and 50's.

Prior to TQM, manufacturers operated on the assumption that there were definite trade-offs among the manufacturing factors: Time, Quality and Cost. If you tried to make improvements in one of these three factors, it could only be done at the expense of the other two. Deming changed that.

The essence of Deming's TQM was twofold: (1) define quality as what are the needs of the customer and put in place a system to identify these needs; and (2) put in place a business-wide process to continuously improve the ability of the business to meet those needs.

The definition of quality was novel and depended on the business being in close touch with its customers, communicating with those customers on a continuous basis and applying this knowledge to provide the business with its strategic direction. This, of course is exactly what Michael Dell did in setting up his PC business to sell directly to customers.

The second part of Deming's TQM was perhaps the more challenging. It involved the entire operations of the business and the entire management structure. In the ideal, each and every function of the business had to be optimized with respect to delivering to the needs of the function's "Customers" within the business. Thus for example, the accounting function had, not only to keep the accounts, but it had to provide accounting information to the parts of the business that needed it to perform their function; and so on down to the factory workers on the manufacturing line and the sales staff selling the product or services to the ultimate customers.

Another objective of the continuous improvement process was to eliminate waste by identifying the causes of waste: poor design, poor processes, and poor workmanship, as early as possible to reduce waste as quickly and cheaply as possible. The TQM system thus could allow the achievement of simultaneous improvements in Quality, what the customer needs; Time, how long it takes to deliver products and processes that meet those needs; and Costs, how much it costs to deliver those products and processes. For the businesses that implemented this system, the result was a significant competitive advantage that, because of the continuous improvement methodology, grew over time.

Deming, and others, tried to get American Companies to adopt this approach but found few takers. However, as a result of several visits to Japan after the Second World War, he found that Japanese manufacturers welcomed him and his ideas with open arms. Their industries had been destroyed during the war and they were anxious to rebuild them in a way that would allow them to compete effectively against world competition. The result is history that is still playing out. Using TQM, or variations of it, Japanese automotive and electronics companies became world leaders;

something still true to this day (2013).

One of the intriguing questions is why American businesses were not much quicker to adopt the TQM way. There are undoubtedly a number of reasons: The methods were revolutionary and perhaps required a huge leap of faith; the methods involved the application of statistical methods pioneered by Deming that had not been taught much in engineering classes of the day and so required a lot of retraining in a field that to this day continues to be difficult to master; but perhaps most of all, it entailed turning the organization chart upside down so that at each level the job of a manager was not to boss his or her employees around but to serve them by doing his or her best to ensure the employees had the required training, to ensure the employees had the required tools and equipment, and to ensure the employees were performing the right jobs. The idea that the job of a manager was to serve the people reporting to him rather than the next level up was revolutionary and contrary to every system of management up to that time. It was going against the grain of managers from the lowest levels to the CEO's, and most found it difficult if not impossible to switch.

(An interesting illustration of the difficulties traditionally managed companies were faced with TQM is the following story told to me, in 1986, by Raymond Royer, then the president of Bombardier, the Canadian world class, transportation equipment company. Royer was a bright, soft- spoken, accountant, who was thoroughly committed to the TQM way. Shortly after Bombardier took over the Irish, aerospace company, Short Brothers PLC ("Shorts"), Royer visited Shorts plant in Belfast, Ireland and was given a full tour that wound up in the executive dining room. On the tour, he was shown four separate dining facilities: the cafeteria for the factory workers, a dining room for the factory managers, a dining room for the office personnel, and finally the executive dining room for the senior officers and directors of the company. During the meal in the executive dining room, Royer passed on the following observation to his hosts, in his lowkey way: "at Bombardier, we all eat in the cafeteria".

Shorts got the message, and a few months later, on his next visit, Royer found everyone eating in the cafeteria, where the food was very good.)

In contrast to most North American companies, the Japanese were starting pretty much from scratch and had a need to be different, so they adopted TQM and used it to produce some of the new world leaders in manufacturing. Of course, American companies caught on and some of the more progressive adopted some or all of the TQM approach. The American car companies were forced to do this in the 1980's to try to meet the Japanese competition. And Michael Dell was one of the high tech leaders who saw the merits of the system and built his company on it.

Dell took the TQM philosophy a step further, using a system called "Integrated Manufacturing". Integrated Manufacturing was a management system that was becoming popular in the 1980's and 1990's. Integrated Manufacturing consisted of identifying the business functions that were critical to the success of the business, focusing all effort and energy on optimizing these functions, and outsourcing everything else to a select group of suppliers who were the best in performing the outsourced functions.

A key element of a successful Integrated Manufacturing approach was communications; both among functions inside the business and with the supply chain supplying the outsourced functions. An Integrated Manufacturing system was designed to keep all internal functions, such as final product assembly, and all suppliers, fully aware on a real time basis of where their products were in the supply and manufacturing chain right through to the ultimate customers. Perhaps more importantly, it would communicate to suppliers where and when their products or services were falling short on Quality, Time of delivery and Costs. This allowed the suppliers to correct any shortcomings as quickly as possible and to minimize rejects and returns. This in turn resulted in meeting more customers' needs more quickly and with substantial cost savings.

One of the consequences of a smoothly operating Integrated Manufacturing system was the achievement of "Just in Time" ("JIT") inventories throughout the business and the businesses of its suppliers. JIT permitted lower working capital requirements and was another contributor to the cost savings possible with the system.

Running these TQM and Integrated Manufacturing systems, Dell became the top performer in the PC business, delivering products that met its customers' needs, on a timely basis and at industry low costs. For example, Dell became known as the best in the PC business at JIT savings. Dell prospered with steadily increasing sales and profits following its IPO in 1988. However, as previously noted, this prosperity was not much reflected in its share price until 1995 when the share price took off and went on a tear of exponential growth for the next 6 years.

The Internet and World Wide Web

What happened in 1995? One thing that happened was the start of widespread commercial use of the Internet. The arrival of the Internet caused a revolution in how PCs were used; they became primarily communication devices rather than work stations; a development that exploded sales of PCs.

Of course, there were networks of personal computers before 1995.Local area networks ("LANs") linked computers in an office so they could communicate with one another. Then Wide Area Networks ("WAN's") proliferated linking computers from office to office across the country. But generally, these were private networks and only proprietary terminals, including PCs, with passwords, could hook into the network.

The Internet was something else. It was, and is, a network of computers that eventually spanned the globe, linking any computer that had networking capability to any other such computer regardless of who owned it or where it was located.

How did this happen and what did it do to the way we use personal computers? Please refer also to Appendix B: A Hierarchy of Network Hardware and Software, and Appendix C: An Early History of the Internet for background material on the Internet.

The Internet[35]

The concept of a global network of computers can arguably be traced to the work of J.C.R. Licklider beginning in 1960.[36] In 1962, Licklider was tasked with the job of creating a network of computers within the U.S. Department of Defense.

In 1969, Robert Taylor headed up a group at the U.S. Department of Defense called the Advanced Research Projects Agency ("ARPA") whose aim was to create a network of computers based on the theories developed by Licklider.

The network was motivated by the cold war of the time to be decentralized to "enable government researchers to communicate and share information

[35] Wikipedia: the Internet
[36] Ibid

across the country in the aftermath of a nuclear attack".[37] The network they came up with used packet switching, "a rapid store-and-forward network design that divides messages up into arbitrary packets, with routing decisions made per packet."[38] This method of routing was far superior to existing methods and allowed faster communications among computers. Packet switching is still the method used by the Internet to this day.

The first ARPANET link using this methodology was established between the University of California, Los Angeles

(UCLA) and the Stanford Research Institute at 22:30 hours on October 29, 1969.

Here is a quote from one of the researchers at that time:

"We set up a telephone connection between us and the guys at SRI (Stanford Research Institute) ...", Kleinrock in LA ... said in an interview:

"We typed the L and we asked on the phone,"

"Do you see the L?"

---

[37] The State of the Net by Peter Clement, p. 9

[38] Wikipedia: History of the Internet

"Yes, we see the L," came the response.

We typed the O, and we asked, "Do you see the O."

"Yes, we see the O."

Then we typed the G, and the system crashed ...

Yet a revolution had begun."[39]

In 1981, the U.S. National Science Foundation created the Computer Science Network expanding access to the ARPANET; and in 1986 "provided access to supercomputer sites for research and education organizations."[40]

It was not until 1992, that the final restrictions on the commercial use of the "Internet" were removed and commercial use started to become widespread.[41] Internet access was originally provided by Internet Information Services, such as CompuServe and America-on- line. By 1995, increasingly internet access was being obtained directly through Internet Service Providers, (ISP's), generally the telephone and cable companies. The cable modem had been developed to allow internet

[39] Gregory Gromov: Roads and Crossroads of Internet History
[40] Wikipedia: History of the Internet.
[41] Ibid

access at faster speeds through the networks of the TV cable companies.[42]

At about this time, the number of computers connected to the Internet began to take off in earnest. In 1994, about 3 million computers were connected to the Internet; in 1995, the number reached about 6.6 million; and by 1997 the number exceeded 16 million. There after the number of "connected" computers grew exponentially.[43]

The Internet Standards (Protocols)

The Internet as we know it today is not a single network but a network of networks that link individual devices (in our case, personal computers) through local area networks and wide area networks to larger networks linked through Routers that pass data from one network to another. The Routers are large computers that act as gateways between networks and"process, filter, forward, route

[42] The State of the Net, by Peter Clement

[43] Ibid

and pass packets from one network to another".[44]

To make this communication system work, there must be a universal set of standards governing all the steps required in the network communication process. Fortunately, these were developed and made available free of charge.

The internet standards are set out in layers, in which each layer deals with a separate aspect of the internet communication process. These layers are complex and not easy for the layman to comprehend. The most general set of standards for networking are set out in the OSI seven-layer Model (the Open Systems Interconnection Reference Model). This Model was developed by the International Organization for Standardization to establish protocols to allow communication over networks connecting computers and other devices regardless of the hardware and software used to establish the network. (See Appendix E for a fuller description of the hierarchy of network standards.)

The following quote from George Gilder's book, TELECOSM, provides a

[44] Internet and Internet Engineering by Daniel Minoli, p. 38

helpful analogy for the seven layer OSI Model of protocols. "For a deceptively familiar example consider a phone call. Pick up a handset and listen for a dial tone (Physical layer); dial up a number (every digit moves the call another (Data) link closer to the destination); listen for the ring (signifying a Network connection and Transport (layer) of signals). Getting someone on the line, you may be said to have completed the first four layers of the OSI stack. Then your hello begins a Session, the choice of English defines the Presentation, the conversation constitutes the Application layer. The hang up ends the session. You may be said to have proceeded through seven layers or so, each dependent on the one before it in the (OSI) stack".

Perhaps the most important set of internet standards is referred to as the TCI/IP protocols. These were first developed in 1982 to provide standardized routing and communications of data across the interconnected network of computers.[45] The TCP/IP model consists of five sets of standards[46], referred to as layers, with each layer

[45] The State of the Net by Peter Clement, p. 10
[46] Data Communications and Networks: by Behrouz Forouzan, 1998, p. 550

dealing with a different aspect of the data communications.

The first two layers, (1) the Physical Layer and (2) the Datalink layer, are not defined in the TCP/IP model; rather it works with all of the standards for these two layers. including the OSI standards (see Appendix E).

(1) The Physical Layer handles the physical connection between the communicating computers through the network cables.[47]

(2) The Data Link layer deals with addressing and similar communication requirements to ensure the link between participating computers

(3) The IP part of the model sets the standards for the Network layer (in the OSI model). This is the layer that governs how the network links two computers and allows the movement of data packets over that network (routing).

(4) TCP corresponds to the Transport layer in the OSI model. It defines how a data message is divided into packets, how a connection is established between two computers, how the packets are

---

[47] Internet and Internet Engineering by Daniel Minoli: p.34, 35

transmitted over the connection, and how they are reassembled into the message at the receiving computer. It also provides reliability and control functions.

(5) The final layer of the five-layer TCP/IP Model includes the functions of the Session, Presentation and Application layers of the OSI Model.

The Session layer "establishes, maintains, and synchronizes dialogs between users or applications;"[48]

The Presentation layer performs translation, encryption/decryption, authentication, and compression of data[49] to provide a reliable flow of data between computers; and

The Application layer handles the details of the particular application, such as E-mail or web browsing.

Each layer depends on the efficacy of the layers above and below it in the scheme of communication standards. For the most part, these network protocols are embedded in the Personal Computer's Operating system and Web browser software.

Another set of protocols or standards that is essential to the universal operation of the Internet is

---

[48] Data Communications and Networks: by Behrouz Forouzan, 1998, p. 525

[49] Ibid

a set of protocols governing the formatting of the data to be transmitted over the Internet. These were developed by Tim Berners-Lee in 1989 "when he conceived the idea for a global "hypertext" system that would facilitate the sharing of information... around the world."[50] Berners-Lee's hypertext system consisted of three protocols: HTTP, Hypertext Transmission Protocol, "allowing Web Browsers to communicate with Web Servers"; HTML, Hypertext markup language, the language in which web pages are written; and URLs, Uniform Resource Locators, that provided the addresses used to identify web pages and other information on the Internet.[51]

The last piece of the Internet standard's puzzle was developed by Marc Andreessen, in the early 1990's. It came to be the Web Browser: "a final piece of software, resident on the user's (computer), through which the elements of HTML code could be viewed."[52]

All of these standards came together, free of charge, to form the

[50] The State of the Internet by Peter Clemente, p, 12
[51] Ibid, p 12
[52] Ibid,p.12

public data communications media, the Internet.

At about the same time, say 1995, this new communications facility set off the building of the physical networks that would carry the network around the world. A group of companies, including Nortel from Canada and Global Crossing, as well as the telephone and cable companies started laying fibre optic cables within cities, across countries and beneath the oceans. A little later, companies began launching communication satellites to enable people with computers in remote areas not serviced by cable to access the internet wirelessly.

The use of the Internet exploded after 1995 and with it the sales of personal computers.

Table of PC sales from Wikipedia. 1996-2011[53]

Year	Annual Sales In millions of units	Year over Year % increase
1996	70.90	
1997	80.60	13.7
1998	92.90	15.3
1999	113.30	21.7
2000	134.70	14.5
2001	128.10	-4.6
2002	132.40	2.7
2003	168.90	10.9
2004	189.00	11.8
2005	218.50	15.3
2006	239.40	9.5
2007	271.20	13.4

[53] Wikipedia: Market Share of Leading PC Vendors. Based on annual press releases from The Gartner Group

2008	302.20	10.9
2009	305.90	1.2
2010	351.00	13.8
2011	352.80	0.5
2012	352.70	-3.5

We can see from the table of unit sales, located on the Wikipedia website and obtained by Wikipedia from surveys carried out annually by the Gartner Group that PC sales grew rapidly from 1996 to 2000. The coincident arrival of the commercial and open Internet in the 1990's and within this service, the ability to communicate by E-mail was a major contributor to the spectacular growth in PC sales at this time.

E-mail

While people found many uses for the Internet over the following years, the predominant use has been E-mail. "Since the mid 1990s, the Internet has had a revolutionary impact on culture and commerce, including the rise of near instant communication by electronic mail, instant messaging, Voice over Internet, "phone calls" "and many other uses.[54]

This is still true today (20132). (See this quote from a 2012 publication: Complete iPad for seniors: "The ability to stay in touch via E-mail from anywhere in the world is one of the main reasons to own an iPAD.") But in the last half of the 1990's, the ubiquitousness, speed and convenience of communicating by Email led to large increases in annual sales of personal computers, the primary device for connecting to the Internet at that time.

E-mail is enabled by a software program embedded in a personal computer's Web Browser. The E-mail software sends the E-mail to E-mail servers that accept, store and forward the messages to the

[54] Wikipedia: History of the Internet

intended recipient(s). Each E-mail message has three components: an Envelope, a message Header that contains the sender's Email address, the recipient's E-mail address and information about the message such as the date and subject heading, and the Message itself. The Envelope provides communication information using the Simple Mail Transfer Protocol ("SMTP") to enable the Internet to route the E-mail from its sender to the intended recipient(s). Once the E-mail has been forwarded and stored in a server, software adhering to either the POP or IMAP protocols allowed the ultimate recipient to retrieve the E-mail on a personal communication device, which in the 1990's was a personal computer. (See also Appendix E)

E-mail greatly raised the bar for speed of communications and the ability to communicate simultaneously with multiple recipients. The general result was a huge increase in convenience, and, most importantly, a huge reduction in cycle times throughout a business and in communications with customers and suppliers. This in turn enabled significant savings in our big three factors: Time, Quality and

Costs, with matching advances in competitive positioning.

Dell revisited

Now let us look at what the Internet did to Dell Computers' business.

As we saw above, Dell's business methodology, "Integrated manufacturing", was highly dependent on good communications, within the company and with Dell's select group of suppliers. The arrival of the Internet and E-mail gave a real boost to the Integrated Manufacturing methodology. It sped up and simplified the communications required by the methodology. This in turn served to increase the competitive advantage provided by it.

As we can see from the table of Unit sales of Personal Computers, sales doubled over five years from 1995 to 2000. Dell with its direct sales based methodology and Integrated Manufacturing management system was a leading participant in these sales increases and in 2001 became the global leader in PC sales, surpassing Compaq, which had been the leader for several years.[55] While Hewlett Packard took over the lead in 2001, after it acquired Compaq,

[55] Wikipedia: sales of personal computers from surveys by Gartner Group

this was a one year phenomena, Dell regained the lead in 2002 and held it for several more years through 2006.

Three Headwinds

In the first decade of the 21st century, the personal computer industry found itself fighting three separate headwinds.

The first of these occurred in 2000 - 2001; the bursting of the Dot.com bubble that eventually took down the whole technology industry and slowed the growth in PC sales.

The second headwind was the financial crisis of 2008 that slowed down all economic activity for an extended period.

The third headwind, and the most important for the PC, was the arrival of the Smartphone and later the iPAD. The Smartphone started to take off in 2004 when Research In Motion, ('RIM") launched its BlackBerry. It took off in earnest in 2007 when Apple introduced the iPhone and Google produced the Android Operating System free of charge. Smartphone manufacturers such as Samsung adopted it to produce a very competitive line of Smartphones. A little later Apple introduced the iPAD tablet that met with immediate and overwhelming success; overwhelming to the PC industry, in the sense that the excitement over electronic communications shifted from the PC to the Smartphone and the iPAD. A key to the success of the Smartphone was and is its use for E-mail.

Three Tailwinds

Nevertheless, in spite of these headwinds PC sales continued to rise, reaching 218 million units in 2005 and 302 million units in 2008. T[56]his continued growth coincides with the development of the World Wide Web ("WWW") as a universal source of "Information", the use of the Internet as a sales channel and the advent of a new phenomenon, "Social Media".

The World Wide Web ("WWW")

The development of the WWW as an information source, providing information from around the world to a connected device anywhere, occurred much more slowly than the take up of E-mail, outlined above, even though both started to be available commercially at about the same time in the 1990's. Undoubtedly this slower development can be attributed to the time taken for people to create the websites and populate them with useful information. For example, Wikipedia, the Internet's free encyclopedia, was started in 2001 and its first articles were

[56] Wikipedia, Ibid

written late in 2001 and 2002. In 2013, Wikipedia contains over 4 million articles.[57]

The World Wide Web can be distinguished from the Internet. "The WWW connects computers that provide information (servers) with computers that ask for it (clients), the communication devices used by individuals like you and me. The Web uses the Internet to make the connection and carry the information."[58] See a video clip describing how the Internet works.[59]

The use of the WWW as an information source required the creation of better and better web browsers. The web browser was a software program designed to go out into the Internet to obtain information and bring it back to the user's PC.

The Web Browser

The first commercially popular Web browser was developed by Mark Andreessen, as noted above, in 1994[60]. Andreessen used the Web Browser to create his company, Netscape, and called the Web

[57] Wikipedia.org/wiki/English Wikipedia
[58] D is for Digital by Brian Kernighan, Chapter 10: The World Wide Web
[59] http://wimp.com/Internetworks
[60] Ibid, Chapter 10

browser, the Netscape Navigator. For years it was the dominant browser, until Microsoft developed the Internet Explorer browser, and effectively used its monopoly of Windows in PC Operating Systems to squeeze out Netscape's Navigator. . In due course, Internet Explorer became the dominant browser in its turn. In the 2000's, Internet Explorer's dominant position has been challenged by other competitive Web browsers; include Firefox and Google's Chrome, Safari for APPLE, and Opera.[61]

Search Engines

Equally important, was the creation of efficient and reliable WWW search engines. "Google", (the author of the most popular search engine,) was founded in 1998, went public in 2004, and had a market capitalization of $200 billion in 2010."[62]

Armed with these tools, we can go to the WWW to obtain information on virtually any subject and at incredible levels of detail.

Today, in 2013, for example, it is possible to obtain from the WWW

[61] D is for Digital by Brian Kernighan, Chapter 6 section 6.5 Applications and Chapter 10; "This book explains how today's computing and communications world operates, from hardware through software to the Internet and the Web" says Amazon Kindle Store.

[62] Ibid

information about a medical procedure that rivals the knowledge of the medical profession; the details of the procedure, how to prepare for it, the risks of the procedure, and in many cases, the success rate of it; all very helpful in deciding whether to go ahead with the procedure or not.

E-commerce

The next Internet growth driver in the 2000's was the development of E-commerce, the use of the Internet as a sales channel. This use of the Internet developed even more slowly than the use of the WWW to access information. With good reason. Most retailers were very leery about using the Internet for sales of products and services, because they were afraid of upsetting their existing channels and relationships. So, for example, a large retail chain with many existing retail stores did not want to create a sales channel that would bypass its existing stores and take away sales that would otherwise have gone to one of them.

There were different options for handling this dilemma. On the one hand, retailers sought to use the Internet as a means of supporting their existing stores. For example, they used the Internet to provide prospective customers with the locations of their stores, with the items for sale in those stores and more recently, with the specific number of each good in any particular store.

Other retailers sought to use the Internet as a sales channel for goods that could be distributed more easily from central warehouses rather than through existing stores and sought to avoid the conflicting internal interests in this way. Some, such as Wal-Mart, did both.

Amazon.com

And then along came Amazon.com that didn't have these conflicts because it didn't have any retail stores.

Amazon.com was incorporated in 1994 and first went on line in 1995. Initially, it used the Internet as a sales channel to sell books and it set up an elaborate supporting infrastructure of warehouses, publisher relationships and book libraries and catalogues to fulfill the sales orders. Later, as we will write, it realized the full potential of e-commerce by providing customers with a device, the Kindle book reader, on which to down load their book orders electronically.

This crucial step eliminated much of the hassle of traditional retailing both for its customers and for itself.

But first let's look at the story of how Amazon was created and developed.

Time magazine made Amazon Founder Jeff Bezos its 1999 Person of the Year, and here is how they tell the story.

In 1994, Bezos, a Princeton graduate in Electrical Engineering and Computer science (reminiscent of Steve Wozniak), was employed by D.E.Shaw, " an unusual firm that prided itself on hiring some of the smartest people in the world and then figuring out what kind of work they might profitably do." Bezos' job with Shaw was to research new business opportunities. In this capacity he became intrigued by the then young communications media, the Internet, and set about trying to determine "what kind of business opportunity might there be here?" Through the facilities at Shaw, Bezos could see that the use of the Internet was expanding in leaps and bounds and he was drawn to the opportunities it was creating for e-commerce. He was looking for a way to use the Internet to create "the most (unique) value for customers; "value" to Bezos would be something customers craved: selection, convenience and low prices, value that would make it compelling for them to switch to a brand new way of buying. In due course, he settled on books. After doing some serious research, Bezos discovered that book wholesalers had their catalogues "in Digital", on CD-ROMs that would be relatively easy to put on line.

Bezos opened the virtual doors of Amazon.com's online store in July 1995, and the company, an instant success, grew like topsy.

But it wasn't until the release of the Amazon e-book reader, the Kindle, in 2007 that Amazon.com realized its full potential in e-commerce, introducing a large improvement in convenience and price for its customers and greatly simplifying its own business model.

I had bought a few books from Amazon.com prior to the introduction of the Kindle, because of the huge catalogue of books it had. ("In 2011, it had a stated library of over 850,000 titles.") But I was an early purchaser of the Kindle and became amazed at how easy it was to buy a book using the Kindle, (perhaps too easy!), and how quickly the "Whispernet" system downloaded the book to the Kindle (and later to other devices and platforms, such as tablet computers).

Amazon.com has added many features to the e-book sales and download system, including reviews, the ability to download a sample of a book, better displays, choice of font size, the ability to add footnotes, as well as ease of payment features. About the only thing you can't do easily is loan your favourite book to a friend.

However, the world of e-commerce and e-books, in particular, has become much more competitive since Amazon obtained the "First Mover" advantage. For example, both Barnes and Noble in the United States and Indigo in Canada have come out with their own e-book readers, the NOOK and the KOBO, respectively, and now achieve a significant portion of their sales on these devices. (other competitors such as eBay have used the internet to create a "many-to-many" approach to selling and have been very successful in so doing.

In response, Amazon.com has added a whole new array of products from luxury goods to doggie pee pads. It acts as a sales channel for other vendors, and has also developed a program, CreateSpace, for authors to self publish their literary efforts through Amazon.com.

Amazon.com has been a stock market success from its IPO in 1997 and has continued to do well for its shareholders throughout the 21 century so far to 2013, in spite of a few hiccups such as the bursting of the Dot.Com bubble in 2000. It continues to have a huge impact on retailing of all kinds.

"The shutting of eight Future shops and seven Best Buy big box stores across the country (Canada) comes as electronics retailers feel the squeeze of the burgeoning business of online retailers such as Amazon.com and Wal-Mart Stores Inc and their cut rate prices. The cuts follow a similar move last March (2012) in the United States, where the parent company said it was closing 50 stores."[63]

---

[63] Toronto Globe and Mail: Report on Business, February 1, 2013

Social Media

A third driver of Internet usage in the 2000's, was the arrival of "Social media".

"Social Media", as the term is used in this essay, refers to the use of the Internet and World Wide Web to enable people and groups of people to interact by exchanging information about themselves, including pictures, and videos, about topics that interest them, and about links to websites of interest to them. This phenomenon has both a positive side and a negative side.

On the positive side, it enables an exchange of information at speeds that could only have been dreamed about just a few years ago; the exchange of information by people around the globe and among groups of people with common interests, some of which are discovered through the social media themselves.

On the negative side, one result is an explosion in the amount of information each of us is exposed to that often leads to information overload and an inability to handle the quantity of information rationally; a second, and perhaps more nefarious negative, is the impact on our individual, or group, privacy of personal information. We will discuss this aspect further below when we look at Facebook since it has tried to address some of the privacy concerns.

One of the important features of Social Media is that it has become a highly interactive media. This is exemplified by the popularity of "Blogs" and "Tweets" as we discuss below. It has facilitated the collaboration of groups of people with common interests at high speeds and from global locations.

As such, it is a very innovative tool that encourages unprecedented innovation in its turn. "Social media differentiates from traditional/industrial media in many aspects such as quality, reach, frequency, usability, immediacy and permanence."[64]

A variety of software enabled tools to facilitate use of this new media have sprung up in the last decade, the first of the 21 century, and more are being developed daily. We propose to look at a few of the more popular tools to give an idea of how the new media is being used. We

[64] Wikipedia on Social Media

will look at Facebook, Blogging, Twitter and Micro blogging, LinkedIn, Skype and YouTube.

The internet usage effects of social media as of 2012 are, according to Nielsen, that internet users continue to spend more time in social media than any other site. At the same time, the total time spent on social media in the U.S. across PC and mobile devices increased by 37 percent to 121 billion minutes in July 2012 compared to 88 billion minutes in July 2011."[65]

Facebook

Facebook is the leading "Social Media" tool in use today, in 2013. It was launched by Mark Zuckerberg in 2004.[66] Facebook remained a private company for several years while its business was becoming established. In 2007, Microsoft bought a small interest in the company that put a market value on the company of about 15 billion dollars. It went public on May 17, 2012 with an Initial Public offering, ("IPO "), that valued the company at $104 billion, although the share

[65] Ibid

[66] Wikipedia on Facebook

price fell back almost immediately and has yet to regain anything like the IPO price since.

The Facebook web page provides a number of tools to assist the user. In general, it provides a place to list the user's personal information, including photographs and videos, and links to websites of interest. The user can also set and limit the names of other users who may access the user's data.

More features are being created almost daily.

Facebook has the same positives and negatives as listed above for "Social Media' generally. It facilitates the sharing of information and communications among individuals and groups, on the one hand, but it raises serious privacy concerns, on the other. The privacy concerns exist in spite of the steps Facebook has taken to try to address these concerns, by for example letting the user limit the people who have access to the user's information or profile. Never-the less this information is in a virtual public domain and the user has no control over how Facebook will use the information or allow others to use it. Facebook derives its revenue almost entirely from advertising and undoubtedly uses the personnel profiles to target advertising.

Other more harmful uses may also be possible.

These days there is increasing concern about "Identity Theft", in which an unrelated unknown party steals your identity information and uses it to take out mortgages on your house, access your bank accounts, etc. without your knowing it until the creditor approaches you to collect his money. Because of this type of concern a lot of people are reluctant to use Facebook and similar social media. On the other hand, many are apparently not concerned about this, or are ignorant of the risks; Facebook is reported to have over 1 billion users, for whom the benefits must outweigh the disadvantages.[67]

Blogging

The term Blogging refers to software that allows aa particular type of publishing on a website. Typically, the author starts a publishing chain by setting up a Blog web page and inviting others to view the webpage, add comments to it, or link to it. The author, or group of authors, can add

[67] Ibid

new material as desired to update the Blog and maintain interest in it. It is thus a means of promoting your own material or interests and seeking collaborative participation by others. The original authors may control the participation of others or not. The subject matter is limited only by the author's imagination and can be as ephemeral as: What will the Internet look like in ten years? Or as down to earth as: Spys are the best apples, aren't they?

The original facility for creating Blogs was a web page publishing software program created by Pyra labs, called the Blogger, and launched in 1999. The company was bought by Google in 2003 and is hosted on Google's website at a sub domain of blogspot.com. Google has since added many features to enhance the original software and make the task of creating and running a Blog easier and more interesting. There are literally millions of Blogs enabling the authors to have a voice on the web and engage with others.[68]

Twitter and Micro blogging

Micro blogging is a type of blog that limits its registered authors and

[68] Google on Blogger

registered visitors to posting text-based messages, called Tweets, of at most 140 characters, to a website. Non-registered visitors can view the chain of tweets but cannot post to it. Micro blogging is used primarily for near instantaneous communications, often to discuss current events. It is most popular among Smartphone users but is also well used by Personal Computer users.

Twitter is the name of the website offering the software service to facilitate micro blogging Tweets. The software It was created by Jack Dorsey in 2006. The service grew rapidly. 1,600,000 Tweets were posted in 2007. As of 2012, there were over 500 million registered users globally, posting over 340 million Tweets a day.

Twitter was reported to be the third highest ranked social networking site in January 2009.[69] It remains a private company, although in 2012 Investment Management Firm BlackRock was able to purchase $80 million of stock from employees, giving the company an implied market value of $9 billion. While financial results are not public,

[69] Wikipedia on Twitter

eMarketer Inc estimates Twitter had advertising revenue of about $ 545 million in the current year.[70]

LinkedIn

LinkedIn is a website that runs software that allows a network of professionals to describe their business to other professionals. It claims to be the "world's largest professional network with millions of members and growing rapidly". By joining LinkedIn and recording your professional profile on its website you can control one of the top search results for your name: and

"Build and maintain a broader network of professionals you can trust.

Find and reconnect with colleagues and classmates. Learn about other companies.

Leverage powerful tools to find and reach the people you need.

Tap into the knowledge of your network.

[70] Toronto Globe and Mail reprint of The Wall Street Week article: Social Media, IPO not a certainty, says Twitter CEO (Dick Costolo): by Shira Ovide; published by the "Globe" on February 6, 2013

Discover new opportunities”.[71]

LinkedIn was founded in 2003 by Reid Hoffman and went public in 2011.59

“We (LinkedIn) recently (January 2013) crossed an important and exciting milestone for the company. LinkedIn now counts over 200 million members as part of our network, with representation in more than 200 countries and territories. We serve our members in 19 languages around the world.”[72]

YouTube

“Founded in February 2005, the YouTube software allows billions of people to discover, watch and share originally- created videos. YouTube provides a forum for people to connect, inform, and inspire others across the globe and acts as a distribution platform for original content creators and advertisers large and small.”[73]

In October of 2009, YouTube announced it had more than 1 billion views per day; and by May of

[71] LinkedIn website
[72] Ibid
[73] YouTube website

2010 it announced it had passed 2 billion views per day.[74]

Skype

Skype is a proprietary Voice over Internet Protocol Service (VoIP) and software application. It allows registered users to communicate from a computer with other parties registered with Skype by voice, video and text messaging. Its main attraction is that Skype to Skype calls from computer to computer travel over the Internet and are free of charges. Skype calls to regular phones or mobile devices bear a small fee. Skype software was developed in 2003 by Estonians Ahti Heinla, Priit Kasesalu and Jaan Tallinn. SkypeIt was acquired by Microsoft in 2011 for $8.5 billion.[75]

"On 19 July 2012, Microsoft announced that Skype users have logged 115 billion minutes of calls over the quarter, up 50% since the last quarter." Today (in 2013) it has over 600 million users.[76]

Other Social media

There are many other Social Media facilities on the Internet. Together,

[74] Ibid

[75] Wikipedia on Skype

[76] Ibid

given their popularity, they constitute an ongoing driver of Personal Computer sales. While much Social Media communications is now accessed by Smartphone and Tablet, much is used on Personal Computers. On February 6, 2013, Information Week cites Yankee Group analyst Chris Walsh as saying that his data suggest that employees still prefer laptops or desktops over tablets for a range of business tasks.[77] It seems this conclusion is also applicable to the use of Social Media, with the likely exception of Twitter, and accounts for a significant part of the growth in PC sales in the last seven years.

While "Social Media" is aimed primarily at building relationships among individuals and groups of like-minded individuals, businesses are increasingly using Social Media sites as new sales and marketing channels.

As Schroeder and Schroeder Inc of Toronto, Canada, a consultancy, point out in their White Paper article on Social Media, these uses are changing the very nature of business. They argue that "the impact of social media is

[77] Information Week

revolutionary; requiring a major change of mindset and approach, in order to keep up with the resulting changes in consumer and business behavior and demands."

Among other things they point out that people place more trust in the opinions of their friends and independent sources than they do in those of vendors. Social media is all about building helpful relationships and so can provide strong sales channels. To make effective use of social media, businesses, argues Schroeder, must retrain their employees to equip them to use the social media effectively and to ensure they provide the right encouragement to their employees to do so.[78]

[78] Schroeder and Schroeder White Paper Series, Social Media, February 2013

The Present – Commodity Years

Personal Computer global unit sales reached 350 million in 2010. But in 2013, we could see that unit sales peaked at 362 million in 2011 and fell off a bit in 2012. In our first book, published in 2013: A Short History of the Personal Computer, we speculated that the PC may have entered a commodity period in which sales would be determined primarily on the basis of price.

One reason for this observation was the entry into the leadership of world PC sales volumes of three Asian PC manufacturers: Lenovo, Asus and Acer. These companies have benefitted from low manufacturing costs and have become three of the top five global PC vendors, along with. Hewlett Packard, having acquired Compaq in 2002, and Dell, the onetime perennial leader.

In the next section, we will see what happened in the next five years.

Intel and Moore's law and diminishing costs

While all the software developments mentioned have been sustaining the high growth path of PC sales, Intel has been leading other chip manufacturers in the evolution of the chip(s) that form(s) the core of the personal computer.

Moore's law states that the number of transistors (on/off electrical switches) on a chip doubles about every two years. "Moore's law inspires Intel Innovation."[79] Intel has succeeded in proving Gordon Moore's prediction continues to come through, in spite of several predictions to the contrary, and has made it happen unbelievably for decades. The current prediction is that Moore's law will run out sometime in 2020 when transistors reach the size of atoms (Quantum Computing) and can't get any smaller.

Intel's continued chip developments, obeying Moore's Law, have repeatedly doubled the speed of chips, halved the power needed to operate them and substantially and continuously lowered their cost. Similar developments have occurred in the devices used to store data. As a result, the cost of producing computers has fallen dramatically, in contrast to the cost of practically everything else.

Unfortunately for Dell and other computer manufacturers the profit

[79] Intel website

margins on personal computers have fallen even faster, making it harder and harder for them to make a buck. This has had a deleterious effect on the stock price of Dell and more recently Hewlett Packard.

The three principal causes of this problem have been: (1) a lengthening of the personal computer replacement cycle, (2) the arrival of the Smartphone in 2004 and the tablet computer in 2009 and (3) The lack of a new Mega App or group of Mega Apps.

(1) Users are not replacing their personal computers as quickly as they once did. The fact is that increasing speed inside the computer is not addressing the primary speed problem of users; the latter lies in the speed of access to the Internet, which has improved slowly in the last few years.

(2) The Smartphone has taken a lot of the action away from the personal computer, principally in the e-mail field, which as we saw earlier was, a primary driver of PC sales from the inception of the public internet in 1995.

(3) There have been no new mega Apps for the PC since the advent of the Social Media Apps.

In 2017, we can see that the peaking in global PC sales has occurred and indeed we are now in a period of precipitous decline in global PC Sales. This is clearly seen in the chart of PC sales running from 2006 to 2017. The chart shows the peak at 365 million in 2011 and a continuous decline to 250 million in 2017, a drop in 6 years of almost a third of the peak number.

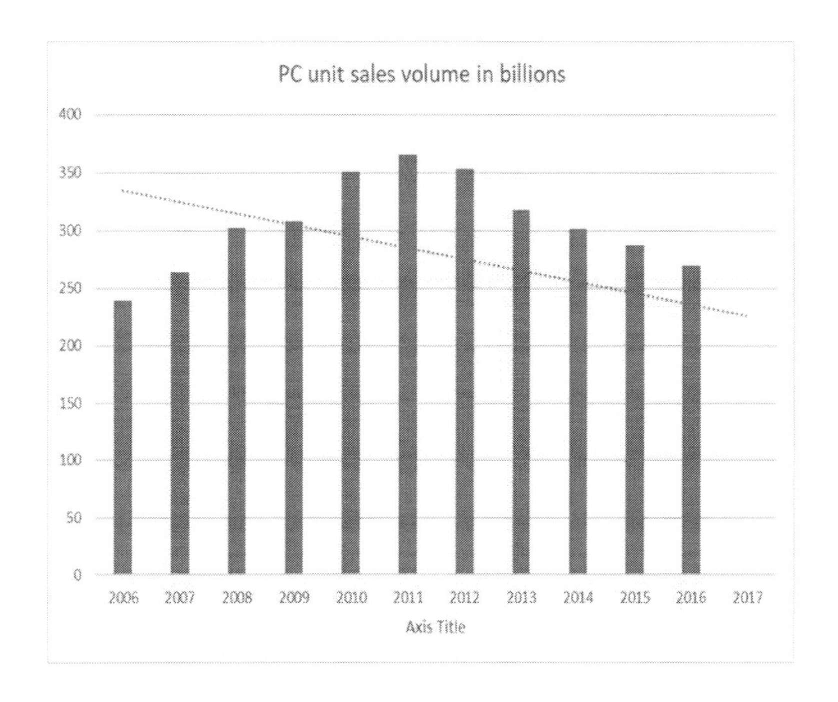

Here we will take a closer look at the two principal reasons for the decline in PC sales and speculate on whether there were other factors that contributed to the decline.

(1) The slowdown in the replacement cycle for PCs.

The serial improvements in the productivity of the PC have not been surrounded by the excitement that

Apple has generated for each release of the new model of its iPhone

Apple has succeeded in convincing its users that each new model represents a significant improvement on the previous model, such as increased picture clarity, improved feel, touch and size, improved multitasking capability, etc., that they must go out and buy the new model. This, by itself has generated successive waves of increased sales of the iPhone, to Apple's profit, and, as the improvements are matched by the other smartphone makers, in a general increase in smartphone sales.

This excitement over new models has not been a feature of PC sales. In part this may be due to the fact that there is not a market leader in PC sales as Apple is for the smartphone. (Improvements in the operation of PCs has been decidedly less dramatic than those of the smartphone but have been more incremental perhaps.) But in part, it may be due to the fact that the improvements in the PC itself have not matched the improvements in the microchip following Moore's law.

Improvements in speed of the microchip are not being matched by improvements in the speed of the

PC. And it is not clear that greater speed of the PC's internal operations is meeting customers' needs. Personally, we find that the speed of our PC has not improved in the recent models, certainly not at the rate of Moore's Law. Similarly, while the price of PCs has dropped, this has not occurred at the rate forecast by Moore's law.

Another factor that might have boosted PC sales would be continuous improvements in ease of use and convenience. This seems to have been the case with smartphones but has not been the case with the PC. If anything, the ease of use of the main PC productivity tools, the Excel spreadsheet, and the Word processor, have not improved at all. If you don't know your way around the tools, it is not easy to figure them out.

And each new iteration requires a new, not easy, learning period. The guidance provided in the Applications themselves may be helpful for computer geeks but are not particularly helpful to the uninitiated user.

Could this disconnection between the measurable improvements in the PC microchips and the operation of the PC itself, be attributable to the lack of similar improvements in the PC Operating Systems? While each release of Windows may have been an improvement on earlier versions, this has not matched the exponential rate of improvements in the microchip, following Moore's Law.

A counter argument is made by Thomas Friedman in his book: Thank You for Being Late. In the book, published in 2016, Friedman presents an analysis of accelerating forces that are reshaping our world in the 21st century, and are impacting each one of us. He describes the accelerating forces and presents suggestions on changes that we as individuals and collectively could make to cope with these accelerating changes.

Friedman identifies three fields in which the rates of innovations are accelerating. The three fields are Computer Technology, Globalization, and the Environment. In the section on accelerations in Computer technology, Friedman discusses the accelerating innovations in Software. He focusses on the accelerations in the development and availability of tools for writing software such as Application Programming Interfaces (API's) that are making it easier and easier to write application software. He also notes that "the tools for writing software are improving at an exponential rate --- but these tools are also enabling more and more people to collaborate to write even more complex software and API codes".[80]

Nevertheless, it is apparent from the current decline in PC sales volume that all the new Apps are not having the collective impact on PC sales that the mega apps, such as Spreadsheets and Word Processors have had. Further, that no recent single App or collection of Apps constitutes a mega App such as the Spreadsheet or the Word Processor. The most recent mega App seems to be the collective Apps that constitute Social Media. This raises the intriguing question; whether there are more PC software mega Apps on the way? We will address this question in the next section of this book.

(2) The competition from the smartphones

In 2004, a Canadian company, Research-in-Motion, released the first smartphone. It was a proprietary system with high level security features that was used for communications by a number of businesses. Because of its

[80] Thomas Friedman's Thank You for Being Late

proprietary nature, while it met with enough success to allow Research-in-Motion to be very profitable, it was never visualized as a consumer product.

In 2007, Apple released the iPhone which was a consumer product with music management and playing capabilities and with a camera, among other things. The iPhone was an instant hit with consumers and sold like hotcakes, to become a best seller and to propel the Smartphone to prominence.

Consumers found it was an ideal communication device because of its size and the fact that it you could take it with you. It had also been made easy to use. The result was the Smartphone surpassed the PC in global unit sales. Today the Smartphone outsells the PC by more than a factor of 5 to 1, with sales hitting over 1.3 billion in 2017. (the enclosed graph tracks Smartphone sales from 2007 to 2017.)

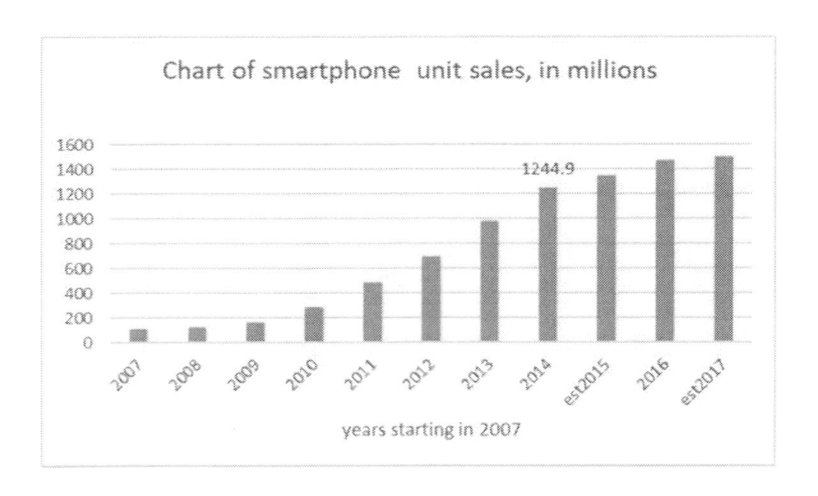

As a result of this success, the Smartphone displaced the PC as the primary digital communication and Internet access device. This in turn took away from the PC one of its primary uses, and undoubtedly has been a major factor in the overall decline in sales of the PC.

The Future of the Personal Computer

The “Next Big Thing” for the PC

The next question is whether there will be a” Next-Big-Thing” for the PC. That remains to be seen. We see four or five possible contenders, maybe more. The first is the Internet of Things, and the second is the ongoing development of Cloud computing. It is conceivable that the PC could become the primary digital device for accessing and managing either or both of these new sources of data and analysis.

We will look first at the Internet of Things.

The Internet of Things

The concept of an Internet of Things has been around for some years. British entrepreneur Kevin Ashton first used the term in 1999 to refer to a global network of radio frequency transmitters embedded in a variety of Things. Today, in 2016, it refers to an internet-connected network of physical devices of all types equipped with embedded sensors and capable of transmitting the sensed information wirelessly to the Internet and from there to a variety of computers, smartphones and the like to be stored, analyzed and used to control the physical devices remotely.[81]

The potential is immense. Already it is estimated by Gartner that there are over 3.8 billion internet-connected “Things” at the end of 2015. It is estimated by ABI Research that over 30 billion “things’ will be wirelessly connected to the Internet and feeding it with all sorts of sensed information. Cisco’s Chairman, John Chambers has forecast that within 5 years there will be 50 billion connected “Things” in a market worth U.S.$ 19 trillion.

In a Toronto Globe and Mail, Report on Business Magazine article, Alec Scott has described the current and potential applications of the Internet of Things. He notes that “these intelligent machines are already altering spheres as diverse as health care, manufacturing, city planning, transportation and power generation, agriculture and

[81] IBM: Can it survive by Peter Farwell

household management". And further, that the devices themselves are "causing macro shifts in how we live and work."

In his Report on Business article, Alec Scott details a long list of Internet of Things applications in play today; in such fields as, Home heat and light controls, Traffic light controls, Manufacturing automation and defect prevention, Robotic cars such as the Tesla, Planes and trains, safety, health and fitness assistance, management of power grids and farming.

Scott also notes that there are a number of obstacles still to be solved; involving issues such as standardization of (computer) languages, security, privacy, preventing hackers from taking subversive control of Things, as well as legal issues about who owns the information, and regulatory issues.

IBM has announced a $3 billion program of R&D focussed on the Internet of Things. One would expect a lot of this would be aimed at developing the infrastructure for the Internet of Things. But already IBM is partnering with many companies targeting specific areas of The Internet of Things. Recent IBM announcements include:

1. IBM and Semtech Corp., a leading supplier of analog and mixed-signal semiconductors, announced a significant advancement in wireless technology, combining IBM software and Semtech hardware to create a system capable of transmitting data up to a distance of 15 km (9 miles), depending on the environment, with significantly improved ease-of-use.

2. IBM and AT&T announced a new global alliance agreement to develop solutions that help support the "Internet of Things." The companies will combine their analytic platforms, cloud, and security technologies with privacy in mind to gain more insights on data collected from machines in a variety of industries.

3. AT&T, Cisco, GE, IBM and Intel formed the Industrial Internet Consortium to improve Integration of the Physical and Digital Worlds. These Technology leaders will establish industry standards for operation of the Internet of Things by identifying requirements for open interoperability standards and defining common architectures to connect smart devices, machines,

people, processes and data being wirelessly transmitted from "Things".

4. IBM, STMicroelectronics and Shaspa announced a collaboration to tap cloud and mobile computing for manufacturers and service providers to provide innovative ways for consumers to manage and interact with their homes' functions and entertainment systems using multiple user interfaces such as voice recognition and physical gestures for a smarter home.

5. IBM and Vodafone announced a collaboration to combine mobile communications and cloud computing for the remote management of 'smart home' appliances.

6. IBM and The Weather Company through WSI, its global B2B division, announced a ground-breaking global strategic alliance to integrate real-time weather insights into business to improve operational performance and decision-making. As part of the alliance, The Weather Company, including WSI, will shift its massive weather data services platform to the IBM Cloud and integrate its data with IBM analytics and cloud services.

7. TransWiseway, a Chinese company, teamed with IBM to design an Internet of Vehicles (IoV) platform to connect millions of trucks as well as tens of millions of devices and sensors from vehicles. Using IBM Internet of Things technologies, the trucks and vehicles are connected to the Internet as well as with each other on this single IoV platform. Built in the cloud, the IoV platform applies analytics to data from mobile devices and sensor data, instantly turning them into valuable information that drivers and authorities can access via the Web or through a mobile app to improve decision making.

8. IBM's Deep Thunder is a high-resolution weather forecasting system that provides customized weather forecasts for business operations. Applications include business areas impacted by weather include insurance, supply chain, agriculture, airlines, and renewable energy.

9. IBM announced a new transportation management solution to help minimize congestion and improve traffic flow for the New Jersey Turnpike Authority. This first of its kind transportation management solution will help

minimize congestion and improve traffic flow for the Garden State.

10. IBM and Eurotech announced that they are contributing software to accelerate and support the development of a new generation of smarter wireless and mobile devices.

The technology, which could become the basis for a new standard of mobile connectivity and interoperability, will be contributed to the Eclipse Foundation open source community.

In a recent report examining technology adoption trends by cities, Forrester Research, Inc., cited IBM for its full set of smart city solution components, making IBM one of only two vendors that is truly a smart city service provider.

Clearly IBM is very actively focused on developing the Internet of Things and has been for some time. But there are big and tough competitors pursuing the same opportunity. These include Cisco, the giant computer net working systems leader, Siemens and other long-time manufacturers of control systems, such as Honeywell International. Still if the forecasters are right this is a huge opportunity and clear leaders have yet to emerge.

This may also be an opportunity for the Personal Computer. The PC could be used as a device that receives all the new data, analyzes it and directs the use of the analysis. But this will require the development of new Application Software to carry out these functions for different industries.

An example of the type of software innovation we have in mind is the Event Stream Processing system offered by SAS.

SAS (Statistical Analysis Systems) was started in 1976 in North Carolina with an assignment to produce a software system that would analyze agricultural information to improve the efficiency of farm management. It grew quickly to help clients in many industries, such as pharmaceuticals, financial institutions, education and government.[82] The rapid growth of SAS over the next few decades, was in part due to the fact its software ran across all platforms, "using the

[82] SAS Website: SAS.com

multivendor architecture for which it is known today".[83]

In response to the development of the Internet of Things, SAS offers a system called Event Stream Processing to receive the mass of data from the Internet of Things, analyze and store the data and release it in real time for use in managing the system of Things producing the data.

SAS describes the opportunities presented by the Internet of Things this way:

"The large array of connected devices (The Internet of Things) is delivering an array of new data from the sensors contained in the Internet of Things."

This data offers the promise of new services, improved efficiency and, possibly, more competitive business models. Connected devices will result in a dramatic shift in how we all interact with computers. First and foremost, computers, computing devices and applications will surround us in an environment where the physical and virtual worlds are constantly connected. Finding a way to process this information – and spot useful events

[83] SAS Website: SAS.com

– will be a distinctive factor in an IOT world."[84]

The SAS Event Stream Processing was designed with this opportunity in mind. SAS lists ways in which its system helps businesses take advantage of the IOT:

"It detects events of interest and triggers appropriate action;

It aggregates information for monitoring;

It polices sensor data to ensure it is valid; and

It permits real time predictive and optimized operations."[85]

An example of the SAS Event Stream System at work, is its use to allow Manheim, the European Automotive remarketer, to transform its business and increase its market share over a two-year operation. Manheim is the largest remarketer of vehicles worldwide, auctioning 10 million vehicles a year

Manheim wanted to improve the way it integrates, analyzes and exploits the mass of data it draws in its business, from car auctions, to

[84] Ibid

[85] SAS literature

enhance the experience of the auto dealers which are its customers. Among other things, it used SAS Enterprise Guide to make it easy to use SAS in a Windows environment.

For its transformation, Manheim needed to quickly analyze and model large volumes of data about the 10 million cars it auctions every year and the thousands of dealers that use its services. It wanted to realign its dealer network to improve performance, so the dealers could see where a particular model was achieving the best price and used the analysis of the mass of data to accomplish this.

The resulting realignment of the dealer network has been a resounding success; a 17% increase in the dealer market over two years and a 15% increase in dealer volume.

Both the general outline of the SAS Event Stream System and the Manheim example suggest several opportunities for use of Personal Computers in the system: Personal computers could be used as the receiver and transmitter of the raw rata from the IOT sensors; Personal computers could be used to carry out real-time analysis of the Data as an integral part of the Event Stream System; and Personal computers could be used as the recipient of the Event Stream analysis to allow monitoring and real time management of the IOT Things. For example, as the controller of the flow of goods through a factory.

The Cloud

Our second choice for a development that could lead to increased use of the PC is referred to as The Cloud.

In his latest book, Thank You for Being Late, published in 2016, Thomas Friedman presents an analysis of accelerating forces that are reshaping our world in the 21st century, and are impacting each one of us. He describes the accelerating forces and presents suggestions on changes that we as individuals and collectively could make to cope with these accelerating changes.

Friedman identifies three fields in which the rates of innovations are accelerating. The three fields are Computer Technology, Globalisation, and the Environment.

The starting point of his analysis is the incredible increases in the number of transistors in a microchip that have occurred over the last four decades, in accordance with Moore's Law. These microchips are used as the brains of Personal computers.

Moore's law states that the power of a microchip will double every 2 years.

Today's chip is 3500 times more powerful, 90,000 times more energy efficient and 60,000 times cheaper. One day it has to stop – no exponential goes on for ever" in our physical world; but not yet.

This progress in chip technology is driving the accelerating power of computers, but, argues Freidman, it is matched by accelerating improvements in four other key components of computers: the memory units, the networking systems, the software applications and the sensors feeding data to the computer. Putting together the improvements in all five of these components has allowed a new service to be offered: The CLOUD, which is taking computing to a new level of capacity.

It is the cloud services that, we suggest, may lead to the next major boost in PC sales.

Freidman argues that the cloud services are so important an advance that it deserves a bigger name and he calls it the Supernova. In his book, he describes it this way.

This technological supernova just keeps releasing energy at an exponentially accelerating rate – because all the critical components are being driven down in cost and up in performance at a Moore's law exponential rate. "And this release of energy is enabling the reshaping of virtually every man-made system that modern society is built on, -and these capabilities are being extended to virtually every person on the planet"".[86]

Freidman puts it this way: "the supernova is creating a release of energy that is amplifying all different forms of power – the power of machines, the power of individual people, of flows of ideas, and of humanity as a whole – to unprecedented levels".81

[86] Thomas Friedman's Thank You for Being Late

Freidman gives a number of examples to illustrate the power of the Supernova. Here are two.

The first is in the field of design. Tom Wojec is with Autodesk, a global leader in 3D design which offers software to design of many different objects from buildings to cars to medical instruments. He headed up a design team tasked with building a prototype dinosaur for the Royal Ontario Museum in Toronto, Canada, from a huge recovered fossil. The project was completed, and the model was very successful; but it took over two years and cost over $500,000.

Some years later, Wujec, dropped in on the museum and saw his dinosaur and wondered whether the same model could be made with modern design software tools. Using a Cloud app called 123D catch, in less than an hour's work, Wujec was able, using a smartphone, to convert photos he took of the original model to produce a digital 3D model that was better than the original. To Wujec this was just an example of how "all industries are becoming computable", and this is allowing designers incredible new power and speed for their work.

A second example concerns Walmart's effort to compete with Amazon in the retail world of e-commerce, taking advantage of the power of the Supernova. In 2011, after failed attempts to create a system that would be competitive with Amazon, Walmart was able to quickly develop a mobile App to be used by its customers to search for a product, to order and pay for its purchase, and to arrange for delivery of the product from a nearby Walmart store, or one of the new Walmart fulfillment centers. To develop this App, Walmart has used the advances, embodied in the Supernova, using Hadoop to handle big distributed date bases, and GitHub to access the immense library of retail software now available.

These are but two of many examples of how the power of The Cloud (Supernova), is being used to create innovations in a wide variety of industries.

It could be that the personal computer will be the main access device for the Cloud and the range of innovations that are being enabled by it. However, the competition from smartphones and tablets and newer devices is fierce. The personal computer may not be able to win this battle.

Advances in Artificial Intelligence

A third possible driver of Personal Computer sales may be advances in Artificial Intelligence, which a Time Special Edition magazine has termed "The Future of Mankind".

In 1956, a small group of mathematicians and scientists met at Dartmouth College to explore the possibilities of "Machine learning". The group is generally credited with inventing the term "Artificial Intelligence, although the seeds of what they were after were established much earlier. As they put it 'every aspect of learning or any other feature of intelligence can in principle be so precisely described that a machine can be made to simulate it".

"Almost all of the hallmarks of our current technological moment-talkative digital personal assistants like Siri and Alexa, genomic-research breakthroughs instantaneous language translations, self driving cars – have at their foundation one key, if broad, thing in common: artificial intelligence, or A.I."[87]

The big AI breakthrough came not through the application of mathematical rules of logic, but by the simulation of a different learning system – Neural networks.

The neural networks in humans and animals reach conclusions by optimizing patterns developed from examples. This is essentially how young children learn before their capacity for deductive reasoning has been developed. They learn from observing repeated examples and using their neural networks to deduce patterns in the examples and to learn that if a certain pattern is repeated often enough, they can generalize a conclusion.

To take a simple example, suppose a child is told repeatedly that eating broccoli is good for it because it is a green food, but eating cauliflower is not. After a sufficient number of iterations, the child will learn that it should eat broccoli, but not cauliflower, and store this conclusion in its memory. The process by which it reaches this conclusion is not deductive logic but the application of its neural

[87] Time Magazine Special Edition on Artificial Intelligence, 2017

networks. The neural networks allow the child to develop a conclusion from the examples given to train it. This is different from the process of deductive reasoning used by adults. Of course, adults use both systems of reasoning to develop their own library of conclusions.

Researchers have learned how to simulate neural networks using special mathematical formulae or recursive algorithms "to learn discrete tasks by identifying patterns in large amount of data."[88].

The whole field of Artificial Intelligence received a boost with the continued increases in power of the computer and the development of access to Big Data sets, facilitated by such developments as the growth of Cloud Computing. These developments allow the accessing of large samples of data, thousands and millions of data, and the processing of these data sets by powerful computers. This capability allows the use of simulations of the brain's neural networks to construct more and more reliable and accurate conclusions, just as the human brain does.

[88] Ibid

The current developments of Artificial Intelligence are proceeding in many research centres around the world; and one of the leaders in these developments is Geoffrey Hinton in Toronto Canada.

As a New York Times article, published in November 2017 put it: "Mr. Hinton, a professor – built a system that could analyze thousands of photos and teach itself to identify common objects like flowers and cars with an accuracy that didn't seem possible." The technique that Hinton used was the simulation of Neural Networks.

Currently, Hinton is exploring an advance using an alternative mathematical technique that he calls a Capsule Network. The neural network that he developed has limitations: "if a neural network is trained on images that show a coffee cup from only one side,- it is unlikely to recognize a coffee cup turned upside down."[89] Hinton's new system will use a different mathematical technique that will operate in three dimensions, not just two. This will allow much more complex conclusions to be developed from masses of

[89] Ibid

examples. Hinton hopes that this new approach will allow computers to "deliver the kinds of autonomous instruments that will improve the efficiency of voice recognition instruments and improve such projects as driverless cars.

Artificial Intelligence is currently expanding its reach to many diverse fields, such as medical diagnosis and treatment optimization, Robotic manufacturing, military applications, astronomy and beyond. The Personal Computer may be the device that people use to access the AI, and perhaps one day to carry out AI simulations of Neural Networks.

Artificial Intelligence, as we know it today, has its limitations. In a December 28, 2017 article in Canada's Financial Post, Claire Boswell notes: "AI systems are now much better than humans at identifying patterns in large amounts of data" – but – "the machines inability to use common sense or generalize means they can't do much beyond the scope of one narrow task." – "Artificial intelligence has a long way to go before it can replace most human employees" – "little progress has been made in replicating general intelligence."

Advances in collection and management of Big Data

Big Data, the term has slightly different meanings for different purposes. For our purposes, it means the masses of digital information or data coming into internet connected users over the internet from multiple types of connected devices. These devices were primarily PC's in the 1990's, but now include smartphones, tablets, web servers, and sensors connecting the "Internet of Things". The amounts of data streaming into users increases daily. Much of this data is unstructured and expressed in natural language. Massive systems of large networked mainframes are needed to deal with this large amount of information. More importantly, are the array of software and services needed to collect this data, store it in a secure and reliable environment and in a manner that can be accessed for analysis.

For purposes of determining the potential impact on sales of Personal Computers, there will be a need for access devices. The personal computer would seem to be ideal for this purpose. The PC has a large screen capable of displaying written words, tables or graphs. It also has increasing amounts of processing power so that it can be used for analysis of data provided from servers and mainframes.

These capabilities would seem to make it the premier access device for extracting valuable information from Big Data sets.

These are four areas of current technological advancement for which the Personal Computer could become the primary access device. This will require the development of the application Software, such as the SAS "Event Stream" system, that can be best used on personal Computers. For this to happen, the personal computer will have to beat very tough competition from the smartphone set and newer devices such as the voice-activated devices being brought onto the market today.

Comparison of Personal Computers and Smartphones

The personal computer has a clear advantage for handling large amounts of information and management of information. The greater processing power and large screen make it the primary device for dealing with spreadsheets, graphs and larger text documents. It will continue to be used by writers and other creators, editors, publishers and readers of text, graphs, and charts. On the other hand, it is not mobile, less convenient to use, harder to use and slower to use, than smartphones and tablets.

Use of either device will require some new application software to enable them to be used as an access device for either information in the Cloud or for data coming from the Internet of Things. The development of this software will likely determine which devices will dominate in this field.

Newer devices

Coming out in 2017 are new forms of competitors for the function of digital access devices. Google has introduced home management devices, Google Home and Google Home Mini, that are entirely voice activated. Apple is bringing out its own home management device,

Home iPod, that is also voice activated. Voice activation systems are improving in leaps and bounds. Some people believe they will replace typing systems entirely. These dedicated devices will pose tough competition for the Personal Computers as digital access devices.

Conclusion

In writing this essay on the history and prospects of the personal computer, we have focused on the fact that at various stages in the life of the PC, the application software that became available for it, played major roles in its growth in sales.

This is not intended to diminish the importance of the ongoing amazing developments in PC hardware. Without these developments none of the rest would have occurred. A summary of the microcomputer developments and operating system developments during the PC's history are set out in Appendix F.

Nevertheless, the demand for PC's and other electronic devices is determined by how useful they are: in the enterprise, how useful they are in improving productivity; and in the home, how useful they are in simplifying tasks, and in extending capabilities, and in both, in cutting cycle times, cutting costs and improving quality.

We have shown: that the VisiCalc spreadsheet software was key to the growth in sales of the APPLE II; that Word Processing software provide a major stimulus for growth in sales of the IBM PC during its first decade; that this was followed by the development of the Internet, including, in particular, the development of universally accessible E-Mail communications and the World Wide Web; and in the 21st century, that the continued growth of the use of the Internet and WWW to access a wealth of information, the development and growth of E-commerce and the explosion of the "Social Media" phenomenon have been major factors behind the continuing increases in PC sales. All of the corporations we have touched on would surely have scored very well on the tests of "Innovative Capacity" to be found at www.corporateinnovationonline.com.

In 2017, we can see from the graph of PC global sales that PC sales have peaked in 2011 and are suffering a precipitous decline. This is attributable to a slowdown in the PC replacement cycle, to the growth in popularity of the Smartphone and the Tablet computers that have taken away a lot of the communication excitement and usage from the PC, and the absence of new mega application software for the PC.

It appears that the traditional heavy users of Desktop PC's and tablets will continue to provide a solid base of PC buyers. Particularly, if the use of these devices continues to be improved, with respect to speed and the power of what can be done using them, and perhaps especially with respect to ease of use. It seems likely that Apple will continue to lead the way in this area.

New uses of the PC will have to be developed if sales are to turn around and grow again. We think that the use of the PC as an access device for the amazing developments now occurring in the four fields we have identified (The Internet of Things, The Cloud, Big Data and Artificial Intelligence), and perhaps other fields, could develop into new uses of the PC.

In most of these fields the competition from the smartphone and tablets will be strong. Competition will also come from newer devices such as the voice activated instruments.

While existing "heavy users" will continue to provide a base level of global sales of PC's, it seems unlikely that sales volume will return to the level reached in 2011. But we expect that the precipitous decline of the years 2012 to 2017 should be halted, and some level of growth resume as the advances we have mentioned encourage new uses of the Personal Computer.

Positive Things to watch For

Improvements in the speed of operation and ease of use of the PC, to match those of the Smartphones, including more intuitive aids for the use of the PC and the mega Apps.

Exciting new applications. We have suggested four areas in which new applications could prosper; The Internet of Things, The Cloud, Artificial Intelligence (AI), and Big Data. The PC could be used as the primary access device for developments in all these areas.

Continual lowering of costs and improvements in speed.

We are deeply indebted to the entrepreneurs, working in both hardware and software, whose vision and hard work created the PC revolution.

References

Wikipedia for various material at www.wikipedia.com

Steve Jobs by Walter Isaacson, 2011,ISBN978-1-4516-4853-9

iWoz: Computer Geek to Cult icon, An autobiography, 2006, ISBN 978-0-39333043-4

D is for Digital by Brian Kernighan, 2012, ISBN-13: 978- 1463733896

Telecosm by George Gilder, 2000, ISBN 0-684-80930-3

Network Essentials by, Dave Kinnaman and LouAnn Ballew, 1999, ISBN 0-070067685-2

How Networks Work by Frank Derfler and Les Freed, 1993, ISBN 1-56276-129-3

The State of The Net, The new Frontier, by Peter Clemente, 1998, ISBN 0-07011979-1

Introduction to Data Communications and Networking by Behrouz Forouzan, 1998, ISBN 0-256-23044-7

Internet & Intranet Engineering, by Daniel Minoli, ISBN 0-07- 042977-4

The Innovators by Walter Isaacson, 2014, ISBN 978-1-4767- 0869-0

Thank You for Being Late by Thomas Friedman, 2016, ISBN 978-0-374-27353-8.

Appendix A

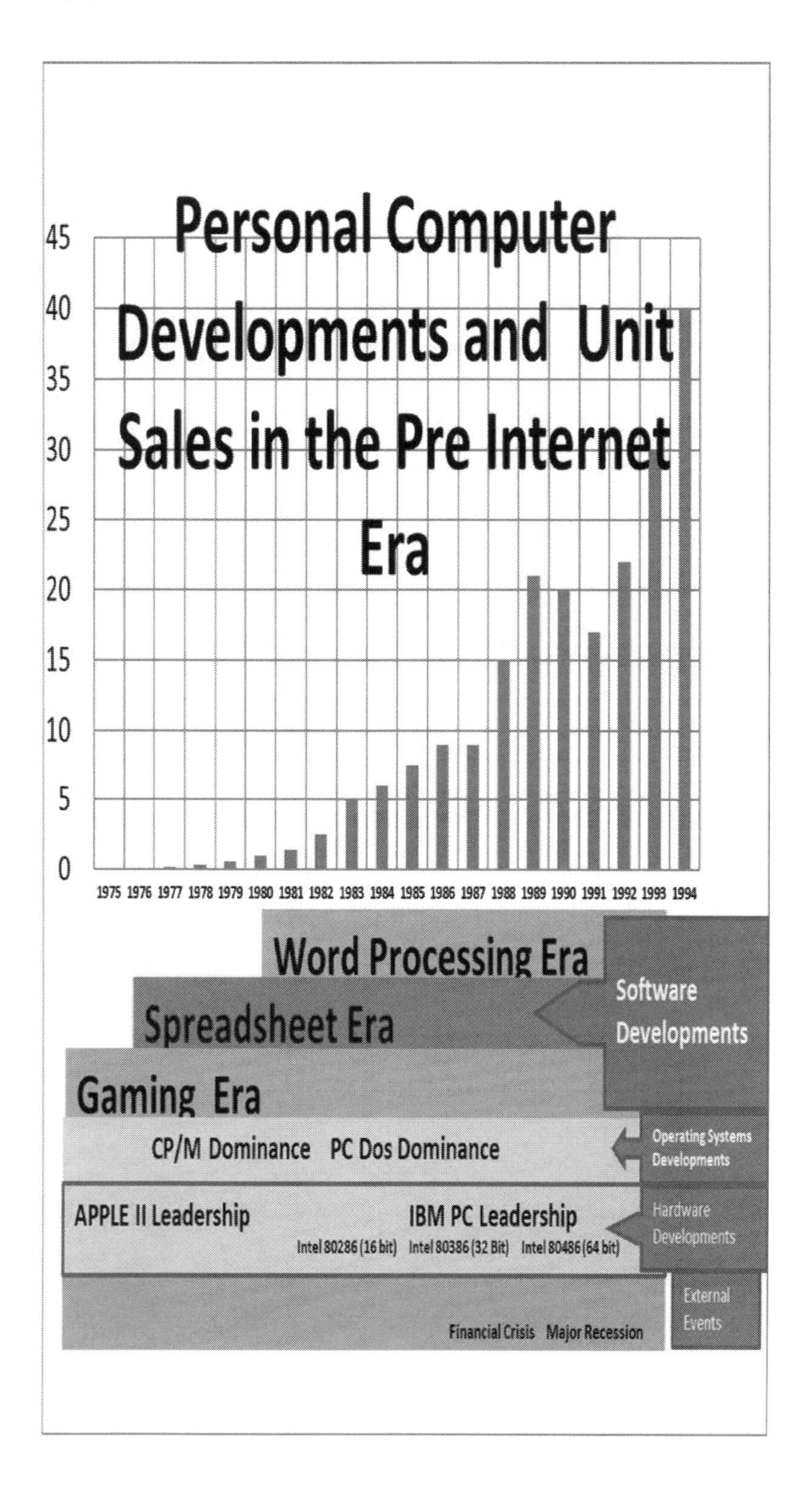
Personal Computer Developments and Unit Sales in the Pre Internet Era
45
40
35
30
25
20
15
10
5
0
1975 1976 1977 1978 1979 1980 1981 1982 1983 1984 1985 1986 1987 1988 1989 1990 1991 1992 1993 1994
Word Processing Era
Spreadsheet Era
Gaming Era
Software Developments
CP/M Dominance
PC Dos Dominance
Operating Systems Developments
APPLE II Leadership
IBM PC Leadership
Intel 80286 (16 bit)
Intel 80386 (32 Bit)
Intel 80486 (64 bit)
Hardware Developments
Financial Crisis
Major Recession
External Events

Appendix B
A Hierarchy of Computer Hardware and Software

Name of Layer Function

Hardware Tangible computing devices, and peripherals.

Transistor An electronic switch that can turn currents on and off, like a light switch.

Circuits Wires that connect the transistors and other components of a computer system, sometimes called buses.

Logic gates The fundamental building block of a computer. An electronic gate made up of a group of transistors, that creates a single output signal (current) based on one or two input signals.

Integrated Groups of logic gates and interconnecting Circuits wiring on a single, small sheet of silicon, or chip.

CPU Central processing unit, that runs

the arithmetic, logical computations,

controls the sequence of operations and moves data within the computer and to and from components.

RAM Electronic memory that can be randomly

and quickly accessed and changed by the CPU, memory that is used to store both active programs and data needed by them. Memory that disappears when power is turned off.

Secondary Internal disk or hard drive, external

Storage storage devices such as DVD's, Optical storage disks, Flash memory devices, that store data and programs permanently for access by the CPU more slowly than RAM.

Input devices Keyboards, mice, microphones, scanners, that provide input to the CPU.

Output devices Monitor screens, printers, etc. that take output from the CPU.

Software Computers are binary processors that

store and process discrete datarepresented by a binary system;

All data and instructions must be represented in binary form, a series of ones and zeros rather than in analogue or continuous form.

This system is usually referred to as Digital for discrete as opposed to continuous data;

Continuous data must be converted to digital form to enable a computer to process it.

Bits The basic representation of data as a binary digit, a number that is either a one or a zero, corresponding to a transistor switch that is either on or off.

Bytes A group of eight bits that allows 256 (2 to the 8th power) different pieces of data to be represented in binary form. Bytes can be combined to represent a wider range of distinct data such as colours or symbols in other languages.

ASCII The American Standard Code for Information Exchange that prescribes what basic data is represented by a particular byte, data such as the letters of the alphabet, both in upper and lower case, the digits, punctuation and mathematical and logical operations. In the beginning, computers were programmed in ASCII (machine language).

Assembly language In time programmers began to use ordinary language to express computer instructions, such as ADD rather than a symbol such as a plus +.

Other programmers developed short routines to carry out standard functions, such as sorting and filing.

The programming language that encompassed these developments was called Assembly Language.

The process of converting from Assembly language to machine language is done by software called Assemblers.

An Assembly language is unique to each type of CPU.

Specific

Programming

Languages New programming languages were developed for specific uses.

Fortran was developed to simplify programming of mathematical calculations.

COBOL was developed for processing business transactions such as inventory management and payrolls.

Basic The Beginners All-purpose Symbolic Instruction Code (BASIC) was developed as a simple, easy to learn and use system.

Basic was the programming language used to program the early PC's.

C, C++ C and C++ were higher level programming languages independent of any particular CPU structure and general in purpose.

They have been developed to make programming easier and safer.

They have been used and are still used to write most broadly used software in plain language.

Programs written in these languages are translated to the Assembly language of a specific CPU structure by a Compiler.

The translation is, in turn, translated into the machine binary code of ones and zeroes by an Assembler program.

Operating systems Operating Systems are the software programs

that manage and control all the operations of a computer.

Their tasks include managing the CPU operations, scheduling and coordinating of different tasks, managing both RAM and secondary memory (the file system), and managing the peripheral devices by Device Drivers.

The first Operating system for the IBM PC was known as CP/M. It was followed by PC DOS (also known as MS DOS)

The most common Operating Systems for Personal Computers today are Windows, UNIX, LINUX and MAC OS.

They provide a standardized platform of services, such as storing and retrieval of data in memory, for Application programs.

Libraries Libraries are software programs that provide generic services, such as date and time references, used in various applications.

Application software Applications are the software programs, such as Word or Excel, that perform specific tasks for the user.

They may be as complex as the Word processors of today, programmed in C++, or as simple as a specific calculation, programmed in Javascript.

The above hierarchy has been compiled with reference to the descriptions in D for Digital by Brian Kernighan with some variations.

Appendix C

Personal Computer Developments and Unit Sales in the Internet Era

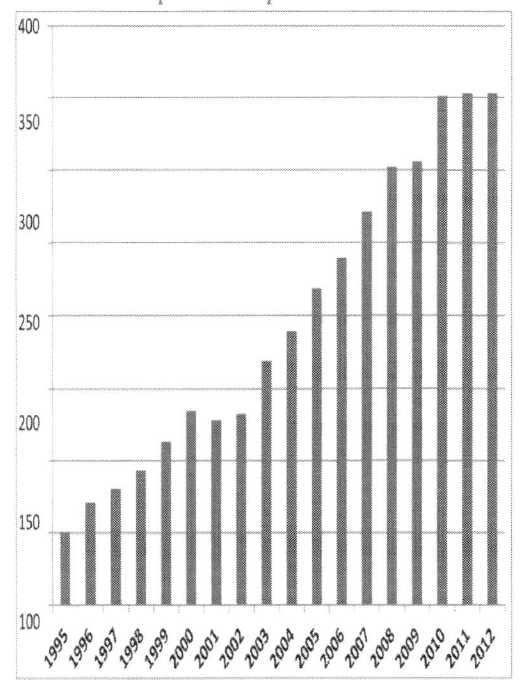

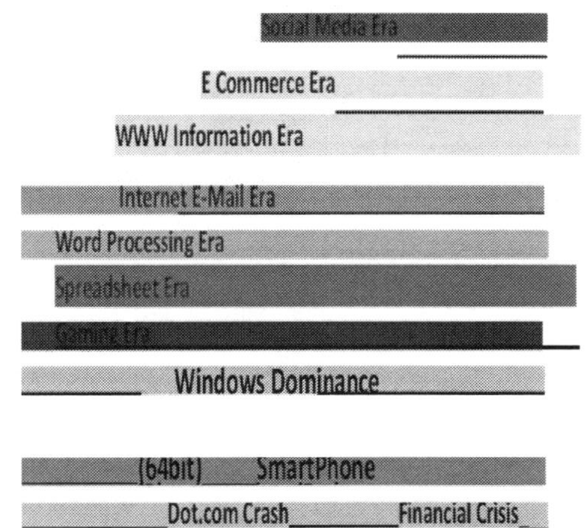

Appendix D

A Hierarchy of Network Hardware and Software

Hardware

Computers In this case, Intelligent communication devices, such as a as a Personal Computer, Smartphone, or Tablet.

Network Interface Card (NIC) A circuit board or chip inside a computer through which the computer connects to a network.

The NIC has one or more Ports from which a network cable will connect the computer to a network.

Network Cable The cable, also called the physical medium, that carries a data signal over the network; it may be a telephone twisted pair line, a cable company's coaxial cable, a fiber optic cable or a wireless network.

Modem **A** network connector that converts digital signals into analogue signals and vice versa.

ISP Cable The cable connecting a network owned and maintained by an Internet Service Provider.

LAN (local Area Network) Generally, a private network connecting computers to Various devices, such as printers, and to a Modem or Router that connects the private network to the internet.

Online Information Server Private companies,such as AOL, CompuServe, Microsoft, and MSN, that provide ISP services.

ISP (Internet

Service Provider)	Today, 2013, typically a telephone or cable company, but also includes the Online Information Servers.
Bridge, Router, Gateway	Network management computers that relay data from one network to another.
Internet Backbone Network	Typically fiber optic cable spanning cities, countries, oceans to link LANs, and other networks, to form the internet.
Websites	A computer connected to the internet that provides information for other internet users.

Software

Operating system	An application program that manages a computer.
Web Browsers	Application programs that allow a user on a connected computer to go out into the internet to communicate with websites on the internet and to obtain data from those websites.
E-mail Systems	Application programs that allow a user on a connected computer to send text messages to and from other connected computers. A standards system called MIME embedded in e-mail software allows the inclusion of graphics, photographs and video in an e-mail

Appendix E

A Hierarchy of Network Standards (Protocols)

Internet Protocols

The Internet is an universal electronic network that permits the transfer of data between computers and between networks of computers

Without these standards, it would be impossible to communicate between the many different types of computers and networks.

The Internet is a "dumb" network; all the "intelligence" is in the software on the computers transferring data over the internet.

This differs from the telephone network in which the telephone terminals are "dumb" and the intelligence is in the network.

Domain Names and Addresses

The names (addresses) used to identify computers and networks on the Internet

Domain Names are usually referred to as "logical names" that use a form of english.

The assignment of Domain Names is managed by a non- profit organization called ICANN, the Internet Corporation for Assigned Names and Numbers.

Domain Name System

The Domain Name System (DNS) sets out a hierarchy of names for computers accessing the Internet.

The top-level names the type of organization or country of the computer,

while lower levels provide increasingly specific names of

networks and computers.

The DNS system provides names used at the application layer.

IP Addresses

IP addresses are 32 or 64 bit sets of hexadecimal (16) numbers that provide the Physical Layer addresses for each network server and computer connected to the Internet. Assignment of IP addresses is also managed by ICANN.

There is a one-to-one correlation between Domain names and IP addresses although they are assigned in slightly different ways.

Routers

Routers connect networks and select the route to get a data packet from one computer through the different networks to the receiving computer, using information in thedata packet's envelope, added at the Network layer of the OSI network protocols.

Routers are continually talking to one another to identify the best routes for

particular data packets and storing the information in their routing tables.

ISO

International Standards Organization formed in 1947 to establish global agreement on communication standards.

OSI Model

Open Systems Interconnection Model, a seven layer set of standards for the design of network systems, that permit communications across all types of computer networks. Each layer sets out standards for a segment of the process of moving data across and between networks, from one computer to another, regardless of the physical equipment, the programming language, platform, operating system, application or user interface (see Network Essentials p 109)

The first three layers define the network, the next four layers define the message and the connection.

Physical Layer

The physical layer sets the standards that allow the different physical components of the network to transmit a bit stream from sender to receiver.

This layer defines the mechanical and electrical specifications for the primary physical connections, the number of connector pins and the function of each pin, between the connecting cables, whether twisted pair, coaxial cable, fiber optic cable or wireless connection.

It permits the transmission of a stream of 0's and 1's over the network in the form of electrical signals.

Datalink layer

This layer defines standards for delivering data packets through each segment of the Network without error. The Datalink Layer is specific to each part of the larger network.It takes a data packet from the Network layer and adds addressing and other control information to form a Frame

The Frame is passed to the Physical layer where it is converted to 0's and 1's for transmission.

Network layer

The network layer sets the standards allowing the data communication packets to be switched (creating temporary connections between physical links) from one link to the next until the packet reaches its intended destination.

It converts logical names and addresses to physical addresses and handles transmission problems such as congestion.

Transport layer

This layer sets the standards that allow a complete data message to be delivered it divides a message into packets for transmission and reassembles them at the destination.

It also provides for error recognition and recovery and flow control.

Session layer

This layer provides standards for controlling a communication session between two application programs on different computer.

It covers user name recognition, user authentication, logins and security.

Presentation Layer

This layer translates data flow into a format usable by the application programs and vice versa.

It also covers data compression and data encryption. It i specific to each type of computer or device.

Application Layer

This top layer enables the user to access the network through familiar application programs by providing standards for a variety of services such as e-mail, remote file access and transfer, shared data base management and other distributed information services.

These descriptions of the seven layers of the OSI Model have been extracted from two sources:

Chapter 3 of Introduction to Data communications and Networking by Behrouz Fourouzan; and

Chapter 9 of Network Essentials by Dave Kinnaman and LouAnn Ballew.

TCP/IP

TCP/IP is a set of protocols developed for transmissions of data over the Internet.

It was originally created by the Advanced Research Project Agency (ARPA) of the U.S. Department of Defense for a private packet-switched network of computers linked by third party lines (the ARPANET).

When the private ARPANET evolved into the public Internet in the 1990's, the evolving standards of the TCP/IP became the protocols for the Internet.

Under the TCP/IP standards the internet operates like a single network linking computers and other communication devices of all types.

Relationship to the 7 layer OSI Model

TCO/IP is a five layer standards model that generally corresponds to the OSI model as follows;

Physical and Datalink Layers

The TCP/IP model does not define standards for the Physical and Datalink layers of OSI. It works with all of the standards, including the OSI standards, for these two layers.

IP (the Internet Protocol)

The IP defines the standards for the network layer that govern how the network linking two computers is established, and how individual packets are formatted and transmitted over that network.

It establishes a "best-efforts delivery service" that provides no error checking or tracking.

Each individual packet from the Transport layer is enclosed at the

IP network layer in an envelope called a datagram. The datagram has two parts:

(1) a header containing addressing and routing information, and

(2) the data packet.

TCP (Transmission Control Protocol)

TCP corresponds to the Transport layer in the OSI model.

It defines how communication data is divided into packets, how a connection over the internet is established between sending and receiving computers, how the packets are sent and how they are reassembled at the receiving computer into the original data. It provides reliability and control functions.

Application Layer

This layer includes the functions of the session, presentation and application layers of the OSI model.

It sets out standards for establishing and managing a session (connection) between application programs on the sending and receiving computers, encompassing user authentication and other security features.

It translates the communication data into a format usable by application programs in the sending and receiving computers, and provides data compression and encryption services.

Finally, this top layer enables the user to access the network through familiar application programs, such as E-mail programs.

The following three protocol sets define the use of the internet for E-mail communications.

SMTP	The Simple Mail Transfer Protocol sets the standards for E-mail communication between computers. Typically, the e-mail program is embedded in a browser, but it may also be a separate program, such as Microsoft's Outlook. The receiving computer is usually a server that is part of a software service, such as Yahoo, that stores the e-mail until it is ready to be retrieved. There are two separate protocols for retrieving e- mails: POP and IMAP.
POP	The Post Office Protocol allows an e-mail program to retrieve E-mails from a server and transfer it to a PC or other access device for reading and storage, while deleting the e-mail on the server.
IMAP	The Internet Message Access Protocol allows the message on the server to be accessed from multiple devices, such as PC's, Smartphones and Tablets, without deleting the message on the server.

Appendix F

The hardware and Operating System software that form the base of the Personal Computer

PC Hardware

The story of PC hardware centres around the work of Intel and the development of the general-purpose Microprocessor that became the core of the Personal Computer. Intel was formed in July of 1968 by the partnership of Robert Noyce and Gordon Moore, with funding provided by Arthur Rock from a list of about a dozen angel investors prepared to back a new technology idea under a team of superior leaders. Andrew Grove joined Noyce and Moore to form the triumvirate that would lead Intel to a revolution.

Both Noyce and Moore were employees of Fairchild Semiconductor for a decade and had worked on various ideas that led to the development of a general-purpose microprocessor, that became the heart of Intel's offerings.

The Intel microprocessor, called the 4004, was conceived as "a general-purpose logic chip that could follow programming instructions"[90]. It was first developed for sale, in 1971, to Busicom to control their stand-alone calculator. Under the agreement with Busicom, Intel retained 'the rights to the new chip and (was) allowed to license it to other companies for purposes

[90] Walter Issacsons's The Innovators, page197

other than making a calculator"[91]. The terms of this deal were critical to the future developments at Intel and ultimately the transfer of power in the PC industry from hardware manufacturers to software developers.

The 4004 was soon followed by the Intel 8080 chip. This was the start of a series of innovative chips (the x86 series).

The Intel 8080 was used in the Altair 8800 microcomputer, "what is considered by many to be the computer"[92]. The Altair 8800 was developed by MITS and was the primary object of interest at the historic 1975 meeting of the Silicon Valley Homebrew Computing club. The first programming language for the machine was Microsoft's founding product, Altair BASIC.[93]

The deal with IBM in 1981

The real takeoff of the PC microprocessor at Intel started with the initiative by IBM to create a true personal computer. This initiative started in 1981, when IBM approached Intel to design and manufacture the microprocessor that would be the core of the new IBM PC. The microprocessor designed by Intel (for this purpose) was the 30826 microprocessor, part of the "86 series" that originated with the 4004. This series was the result of continuous improvements in the power of the Intel

[91] Walter Issacsons's The Innovators, page197

[92] Wikipedia

[93] Ibid, Other early microcomputers marketed in these early days include the Commodore PET, and the TRS-80 from Tandy Corporation. In 1977, the APPLE II was released by Apple, the first Personal Computer to achieve widespread business use, as we have discussed elsewhere in this history.

microprocessor that is now referred to as Moore's law.

Moore's law

Gordon Moore first stated "Moore's Law" in 1965, in response to a request to offer his view of the prospective growth in power of the microchip. He published a paper in which he projected that the power of the Micro chip would double every year for ten years. Specifically, he stated that he expected the number of transistors that could be placed in a microchip would double every year for the next ten years. (Compounding exponential growth of 100% each year). By doubling the number of transistors on a microchip, Intel could cut the cost of producing the microchip by the same factor of two every year; as well the processing speed would double over the same one-year period. In 1975, he revised this projection to state that the number of transistors that could be placed on a microchip would double every two years.

While no exponential growth in the real world can go on forever, the surprising fact is that this doubling every two years, has occurred for the last forty years. The result is that today's chips are 3500 times more powerful, 90,000 times more energy efficient and 60,000 times cheaper than the original.

It is the exponential growth in the power of the PC, following Moore's Law, that has provided the continuous improvement in the PC, that has provided much of the base for the PC's success over 40 years, as described in the main body of this history.

The following list sets out the history of the sequence of microchips, referred to as the "x86" chips, developed and manufactured by Intel that enabled it to fulfill Moore's Law for microchips for over 4 decades:

1971 Intel 4004 is released as first single chip processor (a "computer on a chip"), with 40 transistors.

1972 Intel 8008 the first 8-bit microchip, substantially increasing processing speeds.

1974 Intel 8080 first true general-purpose Microchip with 4500 transistors and ten times the power of the 8008

1975 Intel 8080 used in the Altair 8800

1978 Intel 8086 first 16- bit microchip

1981 Intel 8088 selected by IBM for IBM PC

1982 Intel 80286 released with 134,000 transistors

1986 Intel 80386 is first 32-bit microchip with 275,000 transistors

1989 Intel i860 microchip with over 1,000,000 transistors, to be used in supercomputers

1992 Intel 80486 introduced

1993 Intel Pentium microchip with over 3,100,000 transistors

1994 AMD dispute over licenses is settled by payment, Intel has right to clone 80386 chips

1998 Intel Celeron chip introduced for handheld devices.

1999 Intel Pentium chip introduced

2000 Intel Pentium 4 introduced with 4 million transistors. This microchip had 100,000 times the number of transistors in the original 4004 that had 40 transistors. This number of transistors on a chip approximates the forecast in Moore's Law, representing an exponential compounding at a rate of doubling every two years, or 2 to the power of 17.

2003 Intel chip released for laptops

2005 Apple agrees to use Intel chips in the Mac

2006 Intel produced first quad-core processors

2006 Intel is subject to several patent infringement and anti-competitive behaviour suits.

2006/7 In October 2006, a lawsuit was filed by Transmeta against Intel for patent infringement on computer architecture and power efficiency technologies. The lawsuit was settled in October 2007, with Intel agreeing to pay US$150 million initially and US$20 million per year for the next five years. Both companies agreed to drop lawsuits against each other, while Intel was granted a perpetual nonexclusive license to use current and future patented Transmeta technologies in its chips for 10 years.

In 2009 Intel settled a lawsuit by AMD for anti-competitive behaviour for $1.2 billion.

2008 Intel Atom is released for netbooks

2010 Intel Core family introduced. Intel Core is a line of mid- to-high end consumer, workstation, and enthusiast central processing units(CPU) marketed by Intel.

These processors displaced the existing mid-to-high end Pentium processors of the time, moving the Pentium to the entry level, and bumping the Celeron series of processors to low end. Identical or more capable versions of Core processors are also sold as Xeon processors for the server and workstation markets.

The first Intel Core desktop processor—and typical family member—came from the Conroe iteration, a 65-nm dual-core design brought to market in July 2006, based on the all-new Intel Core microarchitecture with substantial enhancements in micro-architectural efficiency and performance, outperforming the Pentium 4, while operating at drastically lower clock rates. The new substantial bump in microarchitecture came with the introduction of the 45-nm Bloomfield desktop processor in November 2008 on the Nehalem architecture, whose main advantage came from redesigned I/O and memory systems featuring the new Intel QuickPath Interconnect and an integrated memory controller.

Subsequent performance improvements have tended toward making addition rather than profound change, such as adding the Advanced vector management system instruction set extensions, first released on 32 nm in January 2011. Time has also brought improved support for virtualization and a trend toward higher levels of system integration and management functionality through the ongoing evolution of facilities such as Intel active management system.

2012 The Intel Xeon coprocessor is released.

Here is a graph showing the achievement of Moore's Law through 2006.

PC Software

As we look at the Personal Computer market today (2017), we can see that Microsoft's Windows is the dominant Operating System (83%) for Personal Computers and runs on almost all IBM type PC's that form most of the PC market. Competing operating systems are primarily the MacOS (11%) that runs on the Apple Macintosh, the first PC with a Graphic user interface (GUI,) and Linux (2%) that is a free operating system used primarily on mainframe and super computers, but also on some PCs.[94]

We can see that Microsoft has used its dominance in the PC Operating System market to establish dominant positions in some application software markets; for example, Microsoft's Office package that includes the dominant spreadsheet software, Excel, the dominant word processor, Word, and other leading application software, such as PowerPoint presentation software, and Publisher, software to create material for publication. The question is how did Microsoft establish this dominant position? We examine this question in the following paragraphs.

.

PC Operating Software is the software that manages and controls the operation of a PC. "An operating system (OS) manages

[94] Wikipedia on PC Software

computer resources and provides programmers with an interface used to access those resources

An operating system performs basic tasks such as controlling and allocating memory, prioritizing system requests, controlling input and output devices, facilitating networking, and managing files."

This part of the story of the Operating System for Personal Computers can be traced to the formation of the partnership of Bill Gates and Paul Allen. Gates was still in High School at the time. Their interest in the personal computer was piqued by an article in the January 1975 issue of Popular Electronics that showed off the Altair computer, a personal computer kit that allowed computer hobbyists to put together a machine that could be used to do calculations, store them and spit them out at the user's command. It was also used to run computer gaming software. Gate and Allen quickly realized that an Operating System, which controlled the operations of the Personal Computer, could be used to dominate the Personal Computer market, and, as a result, that it would become much more important that the hardware that could be commoditized so that most of the profits in the PC industry would accrue to the software providers[95].

The Altair was managed by an Operating System called BASIC, Beginners All purpose Symbolic Instruction Code.[96] The BASIC Operating System had been created at

[95] Isaacson's The Innovators

[96] Ibid

Dartmouth to allow non-engineers to write software for the PC.

To seize control of the budding Personal Computer Industry, Gates and Allen became very familiar with BASIC. Later, they developed knowledge of more complex Operating Systems such as COBOL.

In 1972, they learned that Intel's 8008 microprocessor, a more powerful upgrade of the 4004, Intel's Initial Computer on a chip, was being used as the core of the upgraded Altair PC.93 They decided to write a BASIC Operating System for the Intel 8008 being used in the Altair PC.

They convinced Ed Roberts, the CEO of MITS, the inventor and marketer of the Altair, that MITS should sell the Altair with the Gates and Allen BASIC Operating System software.

In 1975, Gates and Allen entered into an agreement with Roberts to license the BASIC software to MITS for resale with the Altair. The license was for a ten-year period for $30 per copy. But significantly, under the license, Gates and Allen would retain ownership of the software and MITS was required to make its best efforts to sublicense the software to other PC hardware manufacturers.[97] They foresaw that these provisions would allow Gates and Allen to define the PC market. Thus, they were well prepared when, in 1981, IBM came looking for an Operating System for its new PC.

As we have described in our main text, for some time, IBM had been looking at the

[97] Ibid

creation of a general-purpose microcomputer for a single user that could operate on a standalone basis and allow a user to perform a multitude of tasks. IBM was concerned about cannibalizing its existing computer offerings and so perhaps was deliberately slow in launching its own PC. However, the success of Apple with the Apple II and some other PC manufacturers such as Commodore and Atari, caused IBM to bite the bullet.

Because it had been slow to make the launch decision, IBM set a one-year time limit for the PC development team. This tight deadline caused the IBM PC development team to build their PC on the Intel 8080 series of microprocessors and to look outside for its Operating System.

Jack Sams was the member of IBM's PC development team in charge of software. When the decision to create an operating PC in a year was made, Sams realized he would have to subcontract the development of the Operating System software. He placed an historic call to Bill Gates.[98]

Gates and Allen were well primed for this call, based on their history of writing Operating System software for the Altair and other work in that field. Gates and Allen were aware that Tim Paterson, working at a small company, Seattle Computer Products, had developed an Operating System for the Intel 8008 series of microchips which Paterson labelled QDOS. When the IBM

[98] Ibid

overture came, Paul Allen quickly bought Paterson's QDOS, for any use, for $50,000. Armed with that purchase, Microsoft entered negotiations with IBM to provide an Operating System for the IBM PC, to be known as PC-DOS. While they did not have this software at the start of the negotiation, they were confident they could produce it quickly, by improving the Paterson QDOS system. This they did in 8 months, in time for the launch of the IBM PC.

Gates was adamant that the deal with IBM should be similar to the one he had negotiated earlier with Tom Roberts at MITS for BASIC, that is that Microsoft would retain ownership of the Operating System and beyond that that Microsoft's license with IBM would be non-exclusive, so Microsoft could license the same software under the name MS-DOS to anyone else. Under this deal, Microsoft would keep control of the source code so only it could make changes to the software, thus retaining control over its future development.116 Thus the stage was set for Microsoft's dominance of PC Operating Systems.

The IBM PC was launched in 1981. While it met with immediate success, the takeoff was slow compared to the success of the Apple II, based on the spreadsheet software, VisiCalc.

The IBM PC and its clones were aimed principally at a much larger market, the office market. To appeal to this market, IBM relied principally on word processing software, which was becoming available in the late 1970's and 1980's. At the time of the IBM PC launch in 1981, the leading word processing software was WordStar. But by 1984 it had been overtaken by WordPerfect, which reigned until 1989.

In 1987, IBM initiated an effort to seize back control of the PC market by introducing the PC/2 with features that were an advance over the original IBM PC and a planned new Operating System, being developed with Microsoft. However, this strategy backfired on IBM, and, in 1989, Microsoft came out with its own new Operating System for PCs, "Windows".

Windows incorporated the Graphic User Interface (GUI) and provided a "multitasking environment" for computers. The main components of the GUI were developed at Xerox PARC, but it was Steve Jobs at Apple that realized the significance of it, "how friendly computer screens would become by using metaphors that people already understand such as that of documents on a desk top".[99] The GUI replaced instructions in the form of typed lines of text, used by MS-DOS, with graphic objects on the computer screen that could be manipulated by users to guide the computer, usually by means of a "mouse". Jobs had Apple incorporate the GUI in its offering of the Macintosh line of personal computers starting in 1984. Apple added a number of improvements, such as a fixed menu bar, drop down menus and a trash can.

---

[99] Ibid

Gates was aware of this development at Apple because he was busy writing the Operating System software for the Apple Macintosh. He realized its power and knew that the GUI had been developed at Xerox PARC and was freely available to others to use as they saw fit. So, it became a part of the first Windows Operating System, Windows 1.0. Gates announce the release of Windows in 1985, well after the Macintosh had come out, but the fact that IBM PCs and its clones dominated the market ensured that Windows would be a market leader. Windows 1.0 was not an immediate runaway success.

At about the same time, Microsoft released its word processing software, Word, which was compatible with Windows. We have made the case that the success of Word was a key factor in the eventual success of Windows, but this didn't happen immediately, as businesses were slow to capitalize on the productivity advantages of Word.

Windows 1.0 was the first in a succession of Windows PC Operating Systems that have allowed Microsoft to maintain its dominance in that field to present times (2017). By 1996, Microsoft had obtained a 90% market share in PC Operating Systems.

In 1990, Windows 3.0 was released with an improved program manager and a new icon system that provided a better user interface, a new file manager, support for 16 colors, and greater speed and reliability.[100]

---

[100] Thought magazine

Windows 3.0 was well received and sold over 2 million copies in half a year.[101]

In 1995, Windows 95 was released with greatly improved user friendliness, as a result of the use of an Object-oriented user interface. It also introduced the Start menu and Taskbar to replace the Program manager. It too was well accepted.

In 1998, Windows 98 was introduced with the Internet browser Internet Explorer built in.

In 2000, Windows 2000 based on Microsoft's NT networking technology, was released, and included software updates over the Internet.

In 2001, Windows XP was introduced, with better multimedia support and better performance.

In 2006, Windows Vista was released after a long development period. It contained a redesigned shell and user interface and significant technical changes, with a particular focus on security.[102]

In 2009, Windows 7 was released, incorporating various improvements, such as a better task bar and a home networking system.

In 2012, Windows 8 was introduced. It was an attempt to integrate the Windows PC Operating Systems with those for tablets and smartphones. It also was linked with other Microsoft services such as social media and game offerings. Generally, it was

---

[101] Wikipedia

[102] Wikipedia

not well received by PC users.

In 2015, Windows 10 was released.[103] It was a significant improvement on Windows 8 that included the return of the Start Menu and other measures to improve the user experience.

This lengthy list of Windows versions incorporated enough improvements over the years to keep Windows as the dominant PC Operating System. Surveys suggest that its market share in 2017 was about 85% to 90%.[104]

---

[103] Ibid

[104] Wikipedia

Made in the USA
Columbia, SC
08 March 2018